Praise for *Six Impossible Things*

'[A]n accessible primer on all things quantum … rigorous and chatty.'
  Sunday Times

'Gribbin has inspired generations with his popular science writing, and this [is a] delightful summary of the main contenders for a true interpretation of quantum mechanics. … If you've never puzzled over what our most successful scientific theory means, or even if you have and want to know what the latest thinking is, this new book will bring you up to speed faster than a collapsing wave function.'
  Jim Al-Khalili

'Gribbin gives us a feast of precision and clarity, with a phenomenal amount of information … This could well be the best piece of writing this grand master of British popular science has ever produced.'
  Brian Clegg, popularscience.co.uk

'Elegant and accessible … Highly recommended for students of the sciences and fans of science fiction, as well as for anyone who is curious to understand the strange world of quantum physics.'
  Forbes

Praise for *Seven Pillars of Science*

'Light, to the point and hugely informative. … It packs in the science, tells an intriguing story and is beautifully packaged.'

Brian Clegg, popularscience.co.uk

'[Gribbin] deftly joins the dots to reveal a bigger picture that is even more awe-inspiring than the sum of its parts.'

Physics World

Praise for *Eight Improbable Possibilities*

'We loved this book … deeply thought provoking and a book that we want to share with as many people as possible.'

Irish Tech News

'A fascinating journey into the world of scientific oddities and improbabilities.'

Lily Pagano, *Reaction*

'Gribbin casts a wide net and displays his breadth of knowledge in packing a lot into each chapter … [it] may inspire readers to dig deeper.'

BBC *Sky at Night Magazine*

Impossible, Possible and Improbable

Impossible, Possible and Improbable

Science Stranger Than Fiction

John Gribbin

ICON

This compendium edition published in the UK in 2022
by Icon Books Ltd, Omnibus Business Centre,
39–41 North Road, London N7 9DP
email: info@iconbooks.com
www.iconbooks.com

Previously published in the UK as separate volumes in 2019, 2020 and 2021
by Icon Books Ltd

Sold in the UK, Europe and Asia
by Faber & Faber Ltd, Bloomsbury House,
74–77 Great Russell Street,
London WC1B 3DA or their agents

Distributed in the UK, Europe and Asia
by Grantham Book Services, Trent Road,
Grantham NG31 7XQ

Distributed in Australia and New Zealand
by Allen & Unwin Pty Ltd, PO Box 8500,
83 Alexander Street, Crows Nest, NSW 2065

Distributed in South Africa
by Jonathan Ball, Office B4, The District,
41 Sir Lowry Road, Woodstock 7925

ISBN: 978-178578-882-6

Typeset in Whitman by Marie Doherty

Printed and bound in Great Britain by
Clays Ltd, Elcograf S.p.A.

CONTENTS

ABOUT THE AUTHOR

John Gribbin's numerous bestselling books include *In Search of Schrödinger's Cat*, *The Universe: A Biography*, *13.8: The Quest to Find the True Age of the Universe and the Theory of Everything*, and *Out of the Shadow of a Giant: How Newton Stood on the Shoulders of Hooke and Halley*. He is an Honorary Senior Research Fellow at the University of Sussex, and was described as 'one of the finest and most prolific writers of popular science around' by the *Spectator*.

ACKNOWLEDGEMENTS

I am grateful to the Alfred C. Munger Foundation for financial support while writing this book, and to the University of Sussex for providing a base and research facilities.

As with all my books, Mary Gribbin ensured that I did not stray too far into the thickets of incomprehensibility, and Improbability Eight owes a particular debt to her. The remaining infelicities are all mine.

PREFACE

Half the Answer

In *The Hitchhiker's Guide to the Galaxy*, the answer to 'Life, the Universe, and Everything' is 42. The essays contained in the pages of this book do indeed cover life, the Universe and (more or less) everything, but as there are only 21 of them, the best I can claim is that they provide half the answer to those ultimate questions. Even this modest achievement, however, is more than I had in mind when, back in the days before COVID-19, I combined my fascination with quantum physics and my enjoyment of short-form writing to produce *Six Impossible Things*.

I am not alone in my fascination with quantum physics, so that choice of subject matter needs no explanation. But why try to confront the mystery of the quantum world in short essays rather than the big book that the topic seems to demand? Part of the answer is that I had already tried the 'big book' approach (more than once) and liked the idea of trying something completely different. But the main reason is that the short form offers a particular set of challenges that I enjoy, and a particular kind of satisfaction when it works. Explaining a scientific concept in 3,000 words is often much harder than explaining it in 30,000 words, but harder or not it is a different skill, just as the ability to paint tiny miniatures is a different skill from the ability to paint life-size portraits. I seem to have acquired this skill young – I was sometimes mildly reprimanded during English lessons at school for making my précis too short – and honed it during my time as a journalist, notably with *New Scientist*.

Looking back at some of my early books, it is obvious that they were really a series of *New Scientist*-level essays labelled as 'chapters' and put between covers. Developing from this into 'proper' books with a narrative thread running from the beginning through the middle to the end was a notable achievement, I felt, but I was increasingly attracted to the idea of combining this achievement with writing something shorter to convey the maximum information as briefly (and intelligibly) as possible. The biggest challenge of this kind would be quantum physics – so the idea for *Six Impossible Things* was born. It seemed to work, and as there was a demand for more, the obvious subject to tackle next (as being almost as difficult to understand as quantum physics) was life. As one reviewer commented, the next book should probably have been called *Seven Pillars of Life*, but my journalistic training led me (probably mistakenly, I now realise) to the mild alliteration of *Seven Pillars of Science*. I hope nobody found that too confusing.

The possibility of expanding the series seemed at that point highly improbable, but while re-reading the complete Sherlock Holmes stories during lockdown (actually, listening to the superb narration by Stephen Fry), I was reminded of one of my favourite quotations – 'When you have excluded the impossible, whatever remains, however improbable, must be the truth' – which must also have been a favourite of Conan Doyle, since it appears in slightly different versions in several of the stories. 'What', I mused, 'were the most improbable things we have discovered about the Universe?' To complete the trilogy along the lines I had begun, I had to select *Eight Improbable Possibilities*, taking me halfway to the answer proposed by Douglas Adams. Which seems like a good place to stop and take stock of the story so far.

John Gribbin
February 2022

SIX

IMPOSSIBLE

THINGS

The 'Quanta of Solace'
and the Mysteries of
the Subatomic World

CONTENTS

LIST OF ILLUSTRATIONS

'Alice laughed: "There's no use trying," she said; "one can't believe impossible things."

"I daresay you haven't had much practice," said the Queen. "When I was younger, I always did it for half an hour a day. Why, sometimes I've believed as many as six impossible things before breakfast."'

Alice's Adventures in Wonderland

SOLACE *n.* (*pl.* -**es**) comfort or consolation in a time of great distress.

PREFACE

What's it all About, Alfie?
The Need for Quantum Solace

Quantum physics is strange. At least, it is strange to us, because the rules of the quantum world, which govern the way the world works at the level of atoms and subatomic particles (the behaviour of light and matter, as Richard Feynman put it), are not the rules that we are familiar with – the rules of what we call 'common sense'.

The quantum rules seem to be telling us that a cat can be both alive and dead at the same time, while a particle can be in two places at once. Indeed, that particle is also a wave, and everything in the quantum world can be described entirely in terms of waves, or entirely in terms of particles, whichever you prefer. Erwin Schrödinger found the equations describing the quantum world of waves, Werner Heisenberg found the equations describing the quantum world of particles, and Paul Dirac proved that the two versions of reality are exactly equivalent to one another as descriptions of that quantum world. All of this was clear by the end of the 1920s. But to the great distress of many physicists, let alone ordinary mortals, nobody (then or since) has been able to come up with a common sense explanation of what is going on.

One response to this has been to ignore the problem, in the hope that it will go away. The equations (whichever version you prefer) work if you want to do things like design a laser, explain the structure of DNA, or build a quantum computer. Generations of students have been told, in effect, to 'shut up and calculate' – don't ask what the equations *mean*, just crunch the numbers. This is the equivalent of sticking your fingers in your ears while going 'la-la-la, I can't hear you'. More thoughtful physicists have sought solace in other ways. They have come up with a variety of more or less desperate remedies to 'explain' what is going on in the quantum world.

These remedies, the quanta of solace, are called 'interpretations'. At the level of the equations, none of these interpretations is better than any other, although the interpreters and their followers will each tell you that their own favoured interpretation is the one true faith, and all those who follow other faiths are heretics. On the other hand, none of the interpretations is worse than any of the others, mathematically speaking. Most probably, this means that we are missing something. One day, a glorious new description of the world may be discovered that makes all the same predictions as present-day quantum theory, but also makes sense. Well, at least we can hope.

Meanwhile, I thought it might be worth offering an agnostic overview of some of the main interpretations of quantum physics. All of them are crazy, compared with common sense, and some are more crazy than others, but in this world crazy does not necessarily mean wrong, and being more crazy does not necessarily mean more wrong. I have chosen six examples, the traditional half-dozen, largely in order to justify using the quotation from *Alice*. I have my own views on their relative merits, which I hope I shall not reveal, leaving you to make your own choice – or, indeed, to stick your fingers in your ears while going 'la-la-la, I can't hear you'.

Before offering those interpretations, though, I ought to make it clear just what it is we are trying to interpret. Science often proceeds in fits and starts. In this case, though, it seems appropriate to begin, with another nod to Charles Lutwidge Dodgson, with two fits.

FIT THE FIRST

The Central Mystery

The weirdness of the quantum world is encapsulated in what is formally known as the 'double-slit experiment'. Richard Feynman, who was awarded the Nobel Prize for his contributions to quantum physics, preferred to call it 'the experiment with two holes', and said that it is 'a phenomenon which is impossible, *absolutely* impossible, to explain in any classical way, and which has in it the heart of quantum mechanics. In reality, it contains the *only* mystery ... the basic peculiarities of all quantum mechanics.'* This may come as a surprise to anyone who only remembers the experiment from school physics, where it is used to 'prove' that light is a form of wave.

The school version of the experiment involves a darkened room in which light is shone on to a simple screen – a sheet of card or paper – in which there are two pinholes, or in some versions two narrow parallel slits. Beyond this screen there is a second screen, without any holes. Light from the two holes in the first screen

---

* *Lectures on Physics*, Volume III. In this context, the terms 'quantum physics' and 'quantum mechanics' are interchangeable. 'Classical' physics means everything before relativity and quantum theory.

travels across to the second screen, where it makes a pattern of light and shade. The way light spreads out from the two holes is called diffraction, and the pattern is called an interference pattern, because it is the result of two beams of light, one from each of the two holes, spreading out and interfering with each other. And it exactly matches the pattern you would expect if light is travelling as a form of wave. In some places, the waves add together and make a bright patch on the second screen; in other places the peak of one wave coincides with the trough of the other wave, so they cancel each other out to leave a dark patch. You can see exactly the same kind of interference pattern in the ripples produced on a still pond if you drop two pebbles into it at the same time. One of the distinctive features of this kind of interference is that the brightest patch of light on the second screen is not directly behind either of the two holes, but exactly halfway between those points, just where, if light was actually a stream of particles, you would expect the second screen to be completely dark. If light was made of a stream of particles, you would expect to see a bright patch behind each hole, and darkness in between those patches of light.

So far, so good. This proves that light travels as a wave, as Thomas Young realised at the beginning of the nineteenth century. Unfortunately, at the beginning of the twentieth century another kind of experiment showed light behaving as a stream of particles. These experiments involved electrons being knocked out of a metal surface by a beam of light – the photoelectric effect. When the energy of the ejected electrons was measured, it turned out that for any given colour of light the energy of each electron was always the same. For a bright light there are more electrons ejected, but they

Richard Feynman
Getty Images

still all have the same energy as each other, and this is the same as the energy of each of the smaller number of electrons ejected when the light is dimmed. It was Albert Einstein who explained this in terms of particles of light, what we now call photons – or in his language, quanta of light. The amount of energy carried by a photon depends on the colour of the light, but for any colour all photons have the same energy. As Einstein put it, 'the simplest conception is that a light quantum transfers its entire energy to a single electron'. Turning up the light just provides more photons (light quanta), each with the same energy to give to the electrons. It was for this work, not his theories of relativity, that Einstein was awarded the Nobel Prize. After a hundred years of thinking of light as a wave, physicists had to start thinking of it as a particle – but how could that explain the experiment with two holes?

It got worse. After seeing the wave nature of light cast into doubt by the photoelectric effect experiments, in the 1920s physicists were

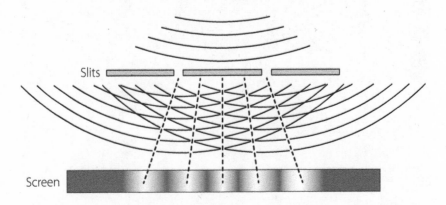

Dotted lines show where waves reinforce, producing bright patches on screen

When light passes through two slits in a screen, waves spread out from each slit to make an interference pattern, like ripples on a pond.

discomfited by evidence that electrons, the archetypal particles of the subatomic world, could behave as waves. The experiments involved beams of electrons being fired through thin sheets of gold foil, between one ten-thousandth and one hundred-thousandth of a millimetre thick, and studied on the other side. The studies showed that the electron beams had been diffracted as they passed through the gaps between the array of atoms in the metal, just like light being diffracted as it passed through the experiment with two holes. George Thomson, who carried out those experiments, received a Nobel Prize for proving that electrons are waves. His father, J.J. Thomson, had received a Nobel Prize for proving that electrons are particles (and was still around to see George get his prize). Both awards were justified. Nothing demonstrates more clearly the weirdness of the quantum world. But this still isn't the whole story.

The puzzle of wave-particle duality, as it became known, lay at the heart of theorising about the meaning of quantum mechanics from the 1920s onward. Much of this theorising about the foundations of quantum mechanics provided the solace for physicists that I discuss later. But the puzzle was brought forth in all its glory in a series of beautiful experiments beginning in the 1970s, so for now I shall skip half a century of solace-seeking to give you the up-to-date facts about the central mystery. If you find what follows hard to accept, remember that as Mark Twain put it, 'truth is stranger than fiction, but it is because Fiction is obliged to stick to possibilities; Truth isn't.'

In 1974, three Italian physicists, Pier Giorgio Merli, Gian Franco Missiroli, and Giulio Pozzi, developed a technique to monitor the equivalent of the experiment with two holes for electrons. Instead of a beam of light, they used a beam of electrons, boiled off from a hot wire, which travelled through a device called an electron biprism. The electrons go into the biprism through a single entrance, but encounter an electric field which splits the beam in two, with half the electrons emerging from one exit, and half emerging from another exit. Then they arrive at a detector screen, like a computer screen, where each electron makes a white spot as it arrives. The spots persist, so as more and more electrons pass through the experiment a pattern builds up on the screen. When a single electron is fired through the biprism, there is a 50:50 chance of it going one way or the other, and it makes a single spot on the screen. When a beam of many electrons is fired through the experiment, they make many overlapping spots on the screen, and these spots combine to make a pattern – the interference pattern expected for waves.

In itself, this is not too alarming. Even if the electrons are particles, there are a lot of them in the beam, and they could be interacting with each other on their way through the experiment to make the interference pattern. After all, water waves make interference patterns, and water is made up of molecules, which can be regarded as particles. But there is more.

The Italian experiment was so precise that individual electrons could be fired through it one at a time, and sent on their way like airliners departing from a busy airport. Like those aircraft, the electrons were widely spaced. The distance from the electron source (actually a bit more sophisticated than a hot wire) to the detector screen was 10 metres, and each electron in the stream did not leave the source until its predecessor had already arrived at its destination. You can (I hope) guess what happened when thousands of electrons were fired one after the other through the experiment to build up a pattern on the detector screen. They made an interference pattern. If the individual particles were acting together to make a pattern in the same sort of way that water molecules interact to make a pattern, then the interaction was taking place across both time and space. This kind of experiment became known as 'single-electron double-slit diffraction'.

When electrons are fired one at a time through the equivalent of the double-slit experiment for light, each electron makes a blob of light on the detector screen. But the blobs build up over time to make an interference pattern, as if they were waves (see image below).

Although the Italian team published these startling results in 1976, they failed to make waves of their own in the world of physics. At that time, few physicists worried about *how* quantum mechanics worked, as long as it did work, in the sense that they could use

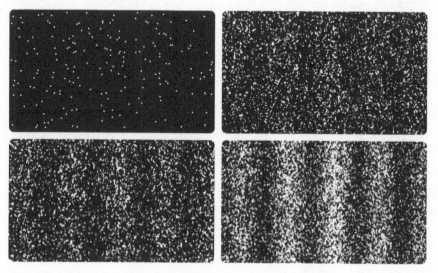

Adapted from A. Tonomura et al., *Am. J. Phys.* (1989)

equations to make calculations and to predict correctly the outcome of experiments. Just how an electron, or a beam of electrons, gets from A to B does not matter to an engineer designing, say, a TV set. You might make an analogy with that vanishing breed of racing drivers who didn't care what went on under the bonnet of their car, but could fling it around the circuit at high speed. The only slightly tongue-in-cheek advice given to students who wanted to know *why* the equations worked was, as I have mentioned, 'shut up and calculate' – that is, use the equations but don't worry about what it all means.

That attitude became increasingly questioned in the 1980s, not least because of the developments which I describe in Fit the Second. So when a Japanese team, headed by Akira Tonomura, carried out similar experiments to those of the Italian pioneers, but using the improved technology of the late 1980s, their results, published in 1989, made a bigger splash. So much so that in 2002, a poll of readers of the journal *Physics World* voted single-electron double-slit diffraction to be 'the most beautiful experiment in physics'. But there was one detail of these experiments that niggled. In the electron biprism experiments there is no physical barrier, like the first screen in the classic double-slit experiment with light, and both routes through the apparatus, both 'channels', are always open. In 2008, Pozzi and another group of colleagues took a step further. They developed an experiment in which electrons could be fired one at a time through two genuine, nano-sized physical slits in a thin screen, to be detected on the other side in the usual way. As expected, the electrons arriving at the detector screen built up an interference pattern. But when the Italian team blocked off one of the slits and carried out another run of the experiment, there was no interference. The pattern on the detector screen was a simple blob directly behind the slit, just as you would expect to be produced by a stream of particles. How does an individual electron travelling alone through the experiment through a hole in a wall 'know' whether there is another hole nearby that it might have gone through, and whether that hole is open or closed, and adjust its subsequent flight path accordingly?

The next step was obvious, in theory, but incredibly difficult in practice. Build an experiment with two holes, on the nano scale, in which the holes could be opened or closed while the electrons were still in flight. Could they be fooled by changing the experimental setup after they had started on their journey? The challenge was taken up by a team based in the USA but headed by Dutch-born Herman Batelaan, who announced their results in 2013. I described their experiment in my Kindle essay 'The Quantum Mystery', and since it involves accurate numbers I cannot improve that description, so here it is again.

The experimenters made two slits in a silicon membrane coated with gold. The membrane was just 100 nanometres 'thick' ('thin' would be a better word), coated with 2 nanometres of gold. Each slit was 62 nanometres wide and 4 micrometres long (a nanometre is a billionth of a metre; a micrometre is a millionth of a metre). The parallel slits were 272 nanometres apart (measuring from the centre of one slit to the centre of the other slit), and, in the crucial new development, a tiny shutter could be slid across the membrane by an automatic mechanism (a piezoelectric actuator) to block one slit or the other.

In the experiment, the electrons passed through the apparatus at a rate of one per second, taking two hours for the pattern to build up on the screen. The whole process was recorded on video. In a related series of runs, the team observed what happened when both slits were open, when one slit was closed, and when the shutter was moved across to block the other slit. As expected, the pattern that built up showed interference when both slits were open, but none for either of the two single-slit options. Once again the electrons 'knew' how many slits were open, on top of all the mysteries revealed (or perhaps I should say confirmed) by the Italian and Japanese experiments. Each electron seemed to 'know' not only what the exact experimental setup was at the time it made its flight through the apparatus, but also what had happened to the electrons that went before it and the ones that would come after it.

Richard Feynman had predicted this would happen, half a century earlier. Drawing on what people knew by then about the behaviour of light, and the discovery of electron waves, he imagined doing the double-slit experiment with electrons. He said in his *Lectures on Physics* that he would describe a thought experiment 'that you should not try to set up' because 'the apparatus would have to be made on an impossibly small scale to show the effects we are interested in'. What was impossible in 1965 proved possible in 2013. It would have delighted Feynman, who among other things was fascinated by nanotechnology. As Batelaan and his colleagues put it, they achieved 'the full realization of Feynman's thought experiment'. It did, indeed, reveal the central mystery of the quantum world laid bare; 'the heart of quantum physics ... the only mystery'. And nobody knows how the world can be like that.

FIT THE SECOND

The Tangled Web

Before moving on, it's important to take away one more lesson from the experiment with two holes. It isn't that things like electrons are seen behaving as both wave and particle at the same time. They seem to travel through the experiment like waves, but they seem to arrive at the detector screen like particles. Sometimes they behave *as if* they were waves, sometimes they behave *as if* they were particles. The *as if* is important. We have no way of knowing what quantum entities 'really are', because we are not quantum entities. We can only make analogies with things we have direct experience of, such as waves and particles. The physicist Arthur Eddington pointed this out in memorable fashion back in 1929. In his book *The Nature of the Physical World*, he said:

> No familiar conceptions can be woven around the electron … something unknown is doing we don't know what. [This] does not sound a particularly illuminating theory. I have read something like it elsewhere —
>
>> The slithy toves
>> Did gyre and gimble in the wabe.

We might, indeed, be better off thinking of slithy toves gyring and gimbling in the experiment with two holes, rather than of electrons behaving as waves and particles. To avoid overkill, I won't be including the 'as if' every time I refer to an event or an entity in the quantum world. But take it as read.

Indeed, 'gyre' might be a better term than the one usually used to denote a fundamental quantum property of electrons, and other 'particles', usually referred to as 'spin'. Spin is a cosy, familiar term, like wave or particle – and just as misleading as either of them. For one thing, the equations tell us that a quantum entity has to rotate twice to get back to where it started, whatever that means in physical terms (and I certainly can't picture it). But spin is a useful property in discussing many quantum phenomena, because it comes in two kinds, which can be thought of as 'up' and 'down'. This simplifies discussions which might otherwise be horrendously complicated.

For example, probability. It was the German physicist Max Born who put the concept of probability, in the context of quantum mechanics, on a secure mathematical footing. But without going into all the mathematics, we can get a feel for its importance using the example of electron spin (or tove gyre, as Eddington might have preferred). It is possible to describe, using the equations of quantum mechanics, an experiment in which an atom emits an electron which travels off through space (this is a real process, called beta decay). In an idealised version of the experiment, the electron has a definite spin. It is either up or down. But there is no way to say in advance what it will be. There is a 50:50 chance of either possibility. If you do the experiment a thousand times, or simultaneously with a thousand atoms, you will find 500 electrons (plus or minus a few, maybe) with spin up, and 500 electrons with spin down. But if you catch a single electron and measure its spin you cannot tell which it will have until you look.

Nothing surprising yet. But Einstein realised that something very surprising is predicted by the equations of the quantum theory for two electrons flying off in opposite directions.* In certain circumstances, a conservation law applies, which says that the electrons must have opposite spin, one up and one down, so that in effect they cancel out. But the equations say that at the time the electrons are emitted from their source, they do not have a definite spin. Each of them exists in what is called a superposition, a mixture of up and down states, and the electron only 'decides' what spin to settle into, in accordance with the rules of probability, when it interacts with something else. The point Einstein seized upon is that if the electrons must have opposite spin, at the moment electron A 'decides' to be spin up, electron B must become spin down, no matter how far apart the two electrons are. He called this 'spooky action at a distance', because at first sight it seems as if the electrons are communicating faster than light, which is forbidden according to the special theory of relativity.

Einstein's idea was developed into a scientific paper, published in 1933, with the help of two colleagues, Boris Podolsky and Nathan Rosen (some might say hindrance, rather than help, since the paper is badly worded and does not bring out the argument clearly). From

---

* Einstein actually discussed this surprise in slightly different terms, but the spin version is easier to get a handle on.

their initials, this is known as the EPR paper, and the point Einstein wanted to make is known as the EPR Paradox, although it isn't really a paradox at all, just a puzzle. In 1935, in a scientific paper which introduced another famous 'paradox', Schrödinger gave the name 'entanglement' to the way two quantum systems seem to be connected by spooky action at a distance. The EPR paper said that quantum theory 'makes the reality of [the properties of the second system] depend on the process of measurement carried out on the first system, which does not disturb the second system in any way. No reasonable definition of reality could be expected to permit this.' Their resolution of the puzzle was that 'we are thus forced to the conclusion that the quantum-mechanical description of physical reality … is not complete'. Einstein thought that there must be some kind of underlying mechanism, known as hidden variables, which would ensure, in this example, that the electrons did not really have a choice about whether to be spin up or spin down while they were flying away from their source, but that everything was predetermined.

Although the publication of the EPR paper provoked fierce debate among the experts, real progress towards an insight into the implications of entanglement was delayed for three decades, largely because one of the most eminent mathematicians of his day, John von Neumann, made a mistake in an influential book on quantum mechanics that he published in 1932, *before* the EPR paper appeared. In that book, von Neumann gave a 'proof' that hidden variables theories could not explain the behaviour of the quantum world – that they were impossible. He was so eminent that everyone believed him, without checking his equations. Well, almost everyone. A young researcher in Germany, Grete Hermann, spotted the flaw in his reasoning and published a paper drawing attention to it in 1935, but only in a philosophy journal not read by physicists and only discovered by them much later. Although, as I shall describe in the second Solace, von Neumann's mistake did not entirely stop people working on 'impossible' hidden variables theories, it wasn't until the mid-1960s that a physicist took von Neumann's argument apart, showed what was wrong with it, and reinvigorated the hidden variables idea. But his revival of hidden variables might not have pleased Einstein, since it also proved that all such theories must include the spooky action at a distance that he abhorred, what is more formally known as non-locality.

That physicist was John Bell, who was taking a break from his work at CERN, the European particle physics laboratory, to work in the USA for a few months on whatever took his fancy. The two papers that emerged from this break from the day job changed what 'everybody knew' about the quantum world as dramatically as anything since the discovery of wave-particle duality. First, Bell explained what was wrong with von Neumann's argument. Then, he showed how it would be possible in principle to design an experiment which would test for the effects of non-locality. More precisely, the experiment would test the assumption of 'local reality'. 'Local' here means that there is no spooky action at a distance – things only influence other things in their locality, defined in terms of how far light can travel in a certain time. 'Reality' is the idea that there is a real world that exists whether or not anyone is looking at it, or measuring it. Because of the probabilistic nature of the quantum world, Bell's proposed experiment would need to

John Bell
Science Photo Library

involve measurements of large numbers of pairs of particles (such as electrons or photons) passing through the apparatus. The hypothetical experiment was designed in such a way that after a large number of runs, two sets of measurements would be produced. If one set of numbers was greater than the other, it would prove that the assumption of local reality is valid. This ratio became known as Bell's Inequality, and the package of ideas as Bell's Theorem. But if the other set of numbers was greater, Bell's Inequality would be violated, which would mean that the assumption of local reality was incorrect. If quantum mechanics is correct, Bell's Inequality must be violated. You can have a real world, with spooky action at a distance. Or you can have locality, at the cost of saying that nothing is real unless it is observed.

Physicists have been down a similar path before, although many physicists themselves do not appreciate it. When, in the seventeenth century, Robert Hooke and Isaac Newton developed their ideas about gravity they realised that the Moon is held in orbit around the Earth by a force that attracts them to each other, and that the planets are held in orbit around the Sun by the same kind of force. They recognised that this was action at a distance. Although neither of them described it as 'spooky', the fact that they did not know how it worked was why Newton famously commented *Hypotheses non fingo* (Latin for 'I make no hypotheses', meaning 'your guess about how gravity works is as good as mine'). He was as baffled by gravitational action at a distance as we are by quantum action at a distance. In the twentieth century, Einstein, with his general theory of relativity, replaced the idea of spooky gravitational action at a distance with the idea of distortions in the fabric of space caused by the presence of matter (although it has to be admitted that some people also find this idea spooky). Perhaps spooky quantum action at a distance will someday be replaced by a less spooky idea by some future Einstein. For experiments have now proved that the phenomenon is real.

Actually carrying out a Bell-type experiment involves technology beyond what was available in the mid-1960s, and Bell did not expect to see the experiment done. But by the early 1980s experiments had been carried out (using photons, rather than electrons) which proved that Bell's Inequality is violated. Many more such experiments, with increasing technical sophistication, have confirmed this since. Local reality is not a valid description of the world; in John Bell's own words, spoken at a meeting in Geneva in 1990, 'I don't know of any conception of locality which works with quantum mechanics. So I think we're stuck with non-locality.' Einstein may have felt that 'no reasonable definition of reality' could allow this, but the conclusion must be that reality is, in his terms, unreasonable. But the most impressive feature of all this is often overlooked. Although the jumping-off point for Bell's Theorem was an attempt to understand quantum physics, and those words were spoken at a quantum physics meeting, these results do not apply only to quantum physics. They apply to the world – the Universe. Whether or not you think that quantum physics might one day be replaced as a description of how the world works, this will not change things. The experiments show that local reality does not apply to the Universe. Whether you choose to find solace in keeping reality and accepting non-locality, or in keeping locality and rejecting reality, is a matter of personal preference, as we shall see. But you can't have both (although you *could* have neither, if you really want to make your brain hurt). Before we seek solace for our aching brains, though, it its worth bringing the story of entanglement up to date, since it has significant practical applications.

Those applications involve a phenomenon known as quantum teleportation. It rests on the now experimentally proven fact that if two quantum entities, such as two photons, are entangled then no matter how far apart they are, what happens to one of them affects the other. In effect, they are separate parts of a single quantum entity. This cannot be used to convey information faster than the speed of light, because what happens to each particle involves probability and randomness. If one photon is tweaked into a random quantum state, the other one is simultaneously tweaked into another quantum state. But anyone watching the second photon only sees a random change obeying the rules of probability. In order for this change to convey information, whoever tweaked the first photon has to send a message by conventional means (slower than light) to tell the second experimenter what is going on. But by tweaking one photon in a certain way, it is possible to change the second photon into an exact copy (sometimes called a clone) of the first photon, while the state of the first photon is scrambled up. In effect, the first photon has been teleported to the location of the second photon. But since the state of the first photon is scrambled, this is not duplication. Once again, the process has to be completed by sending information by a sub-lightspeed process. The teleportation conveys information, but it requires both a 'quantum channel' and a 'classical channel'.

A huge research effort has gone into developing such systems, primarily because the technique offers the prospect of producing uncrackable codes, which would be immensely valuable both to industry and to governments. The essential point is that if any eavesdropper tried to listen in on the quantum channel this would scramble the data, making it useless and revealing the interference. It doesn't matter if the eavesdropper reads the classical channel – as quantum cryptographers point out, it could be printed in the newspapers or published on social media for all the good it would do the eavesdroppers. You need both channels to unlock the coded information. And entanglement is also involved in the development of quantum computers, a topic that is often in the headlines these days. The vision of the researchers is of a totally secure quantum internet, using quantum computation, entanglement, and teleportation to share information utterly securely.

Experiments of this kind have now moved out of the laboratory and into the world at large – and beyond. In 2012, a Chinese team teleported quantum information in this way across the Qinghai Lake, a distance of 97 km. The same year, a European team teleported photons across 143 km, between the islands of La Palma and Tenerife in the Canaries. Both experiments, as an aside, confirmed the violation of Bell's Inequality, something now taken as much for granted by physicists as the fact that apples fall downward from trees.

The Canary Islands experiment involved ground stations on mountains about 2,400 metres above sea level, where the thin air reduces atmospheric interference. But the air is even thinner higher up, and less than 143 km straight up from La Palma takes us to the edge of space. In 2016, China launched the Micius satellite (named after a Chinese philosopher of ancient times), from which beams of entangled pairs of photons were sent to separate receiving stations high in the mountains of Tibet and 1,200 km apart. The satellite was moving at nearly 8 km per second during the experiment, but kept the photon beams on target. To nobody's surprise, but in a triumph of technology, the behaviour of the photons confirmed predictions in line with Bell's Theorem. Although the experiment only operates at night, because sunlight dazzles the detectors, and the success rate of 'recovering' the photons at the ground was only about one in every six million sent from the satellite (fortunately, photons are cheap), there are already plans for a family of satellites with stronger beams that could be detected even in daytime, providing the basis for a quantum communication network, and to teleport photons up to the satellite from the ground. There will probably be more progress, and more headlines, by the time you read this. But while the technologists may continue to 'shut up and calculate', still the physicists cannot agree on what it all means – *why* the world is the way it is.

It's time to look in more detail at some of the ways in which they seek solace. But to bring us back down to Earth, think again about the experiment with two holes. In the experiment, each electron seems to 'know' how many holes are open, and where it is going. Does entanglement – spooky action at a distance – come into the story here as well? If a pair of photons flying in opposite directions are in effect part of a single quantum system, might we regard the whole double-slit experiment and the electron – all of the elec- trons? – as part of a single quantum system? Maybe the electron knows which holes are open because the state of the holes is also part of the state of the electron. But even the notion of entanglement still lay in the future when physicists first sought solace in an interpretation of quantum mechanics which became the standard view for decades.

The Not So Wonderful
Copenhagen Interpretation

The interpretation of quantum physics that became the standard way of looking at things for decades is based on the idea of waves – and on largely forgetting the caveat 'as if'. In the 1920s, physicists already knew that the quantum world could be described in either of two mathematical ways. One involved waves, summed up in the Schrödinger equation. The other involved pure numbers, in the form of arrays called matrices, developed from the work of Werner Heisenberg and Paul Dirac. They gave the same answers, so it was a matter of choice which one to work with; and since most physicists already had some familiarity with wave equations, that was what they chose. In any quantum calculations, however, what you calculate is the relationship between two states of a system, where the system may be an electron, the experiment with two holes, or (in principle) the entire Universe – or anything in between the electron and the Universe. If you have a set of parameters describing the system in state A, you can calculate the probability that it will be

Niels Bohr
Getty Images

in state B after a certain time. But there is nothing which tells you what is going on in between.

The archetypal example is an electron in an atom. Electrons can, for some calculations, be thought of as if (that caveat) they are in orbits which correspond to different amounts of energy. When an atom emits energy in the form of light, an electron disappears from one orbit and appears in another orbit closer to the nucleus of the atom. When an atom absorbs light, an electron disappears from one orbit and appears in one further out from the nucleus of the atom. But it does not *move* from one orbit to the other. First it is here, then it is there. This is known as a quantum jump (or a quantum leap*). Schrödinger intended his wave mechanics to explain what happens during the leap, but it didn't, and he said: 'If all this damned quantum jumping were really here to stay, I should be sorry I ever got involved with quantum theory.' Alas for Schrödinger, it was, and is, here to stay. The matrix approach is more honest, since it does not pretend to try to tell us what is happening between state A and state B, but it provides less solace than the Schrödinger equation.

What was for decades the standard way of looking at the quantum world became known as the Copenhagen Interpretation, because it was vigorously promoted by Niels Bohr, a forceful personality who was based in that city. This name (actually given to the package of ideas by Werner Heisenberg) caused considerable irritation to Max Born, who was not a member of Bohr's team, and did not work in Copenhagen, but whose ideas about probability were an integral

---

* Contrary to what advertisers think, a quantum leap is a very small change made at random.

part of the interpretation. Bohr so dominated any discussions about quantum physics at the end of the 1920s that as well as getting his home town recognised in this way he dissed an alternative, completely viable interpretation of quantum mechanics so thoroughly that it was neglected for two decades. I shall present it as Solace 2.

Bohr was essentially a pragmatist who was happy to stick together different bits and pieces of ideas to make a working package without worrying too much about what it all meant. As a result, there is no straightforward, definitive statement of what the Copenhagen Interpretation is, although Bohr came close to such a revelation in a talk he gave at Como, in Italy, in 1927 – long before the interpretation got its name. The conference at which that talk was given was a landmark moment in physics, because it marked the point where physicists were presented with the tools they would require in order to 'shut up and calculate', applying quantum mechanics to the solutions of practical problems involving atoms and molecules (for example, chemistry, lasers, and molecular biology) without having to think about the fundamentals of what it all meant.

Bohr's pragmatic approach extended to his interpretation. He said that we do not know anything except for the outcomes of experiments. These outcomes depend on what the experiments are designed to measure – on the questions we choose to ask of the quantum world (of nature). These questions are coloured by our everyday experiences of the world, on a scale much larger than atoms and other quantum entities. So we may guess that electrons are particles, and build an experiment designed to test this in an obvious way by measuring the momentum of an electron, thinking of the electron as a tiny pool ball. When we do so, lo and behold, the experiment measures the momentum of the electron, confirming our notion that electrons are particles. But a friend of ours has a different idea. She thinks that electrons are waves, and designs an experiment to measure the wavelength of an electron. Lo and behold, her experiment gives a measurement of the wavelength, confirming her notion that electrons are waves. So what, says Bohr. Just because the electron behaves *as if* it were a particle when you are looking for particles, or *as if* it were a wave when you are looking for waves, doesn't mean that it is either, let alone both. What you see is what you get, and what you see depends on what you chose to look for. It is meaningless, according to the Copenhagen Interpretation, to ask what quantum entities such as electrons and atoms are, or what they are doing, when nobody is measuring them – looking at them, if you like.

So far, so pragmatic, and nothing really too alarming. But Bohr quickly takes us into muddy waters. This is where probability comes in. When Schrödinger came up with his wave equation, he thought of it as being a literal description of an electron (or other quantum entity; electrons are the simplest example to use for illustration). To him, an electron *was* a wave. But Bohr took Schrödinger's ball and ran off with it, combining it with Born's ideas on the role of probability to produce a bizarre and troubling concoction which worked (and still works), as far as quantum calculating was concerned, but makes your head hurt when you stop to think about it. The equation that Schrödinger gave us is, on this new picture, to be thought of as a 'probability wave', and the chance of finding an electron at any location is determined by 'the square of the wave function', essentially by multiplying the equation that describes the wave by itself, at any point. When we make a measurement, or observe a quantum entity, the wave function 'collapses' to a point, determined by the probabilities. But although some locations are more likely than others, in principle the electron could appear anywhere that the wave function has spread to. A very simple example highlights the oddity of this behaviour.

Think of a single electron trapped in a box. The probability wave spreads out to fill up the box evenly, meaning that there is an equal chance of finding the electron at any location inside the box. Now drop a partition down the middle of the box. Common sense tells us that the electron must now be trapped in one half of the box. But the Copenhagen Interpretation (CI) says that the probability wave still fills each half of the box and the electron might with equal probability be found on either side of the partition. Now divide the box in two down the centre of the partition. Keep one half-box in your laboratory, and put the other one on a rocket which takes it to Mars. Still, according to Bohr, there is a 50:50 chance of the electron popping up in the box in the lab or the one on Mars. Now open the box in your lab. Either you find an electron, or you don't. But either way, the wave function has collapsed. If your box is empty, the electron is on Mars; if you have the electron, the other box is empty. This is *not* the same as saying that the electron 'always was' in one half of the box or the other; the CI insists that the collapse only happens when the contents of the box in the lab are examined. This is the kernel of the idea behind the EPR 'paradox', and Schrödinger's famous puzzle involving a dead-and-alive cat. But before going into that story, I want to look at how the Copenhagen Interpretation 'explains' the experiment with two holes.

Erwin Schrödinger
Getty Images

Werner Heisenberg
Getty Images

According to the CI, which I was taught as a student, and which too many students are still taught today, as 'the' way to 'understand' quantum mechanics, an electron is emitted from a source – an electron gun – on one side of the experiment as a particle. It immediately dissolves into a 'probability wave' which spreads through the experiment and heads towards the detector screen on the other side. This wave passes through however many holes are open, interfering with itself or not as appropriate, and arrives at the detector as a pattern of probabilities, higher in some places and lower in others, spread across the screen. At that instant, the wave 'collapses' and turns back into a particle, whose position on the screen is chosen at random, but in accordance with the probabilities. This is called 'the collapse of the wave function'. The electron travels as a wave but arrives as a particle.

The wave, however, carries more than just probabilities. If the quantum entity has a choice of states it can be in, such as an electron which may be spin up or spin down, both states are somehow included in the wave function, the situation called a 'superposition of states', and the state the entity settles into at the point of detection, or interaction with another entity, is also determined at the moment the wave function collapses. In a lecture at the University of St Andrews in 1955, Werner Heisenberg said 'the transition from the "possible" to the "actual" takes place during the act of observation'.

This works as a method of calculating quantum behaviour, as if things like electrons really did behave like this. But it also poses many puzzles. One of the *most* puzzling is a so-called 'delayed choice' experiment, dreamed up by the physicist John Wheeler.

He started from the fact that when photons are fired one at a time through the experiment with two holes they still build up an interference pattern on the detector screen. But according to the CI, if a device is placed between the two holes and the detector screen to monitor which hole the photon goes through, the interference pattern will vanish, showing that each photon really did go through just one of the holes. The 'delayed choice' comes in because we can decide whether or not to monitor the photons *after* they have passed the screen with two holes. Of course, human reactions are not fast enough to do this. But experiments have been carried out with automatic monitoring devices to do exactly this, switching the monitors on or off after the photons have passed the holes. They show that the interference pattern does indeed disappear when the photons are monitored, meaning that each photon (or the probability wave) only goes through one hole – even though the decision to monitor the photon was made only after it had passed the holes.

Wheeler pointed out that you can imagine a similar experiment on a literally cosmic scale. In a phenomenon known as gravitational lensing, light from a distant object, such as a quasar, is focused by the gravity of an intervening object, such as a galaxy, so that it follows two (or more) paths around the gravitational lens. This makes two images of the object in detectors here on Earth. In principle, instead of making those two images it would be possible to merge the light coming different ways around the gravitational lens to make an interference pattern, caused by waves going both ways round the lens. A cosmic version of the experiment with two holes. But then we could monitor the photons before they get a chance to make the interference pattern to see which way round the lens they have come. In that case, according to the results of the laboratory-scale experiments, the interference pattern would disappear. The quasar might be 10 billion light years away, the galaxy acting as a gravitational lens might be 5 billion light years away. But according to everything we know from experiment, what the photons were doing billions of years ago and billions of light years away is affected by what we choose to measure here and now. What is going on? As Wheeler himself put it, 'the Copenhagen Interpretation commands us not to ask such things'.* Not so wonderful, then.

In essence, the Copenhagen Interpretation says that a quantum entity does not have a certain property – any property – until it is measured. Which raises all kinds of questions about what constitutes a measurement. Does human intelligence have to be involved? Is the Moon there if nobody is looking at it? Does the Universe only exist because human beings are intelligent enough to notice it? Or does the interaction of a quantum entity with a detector count as a measurement? Or where in between those extremes do you find the boundary between the quantum world and the 'classical' world of good old Newtonian physics? It was this kind of concern that led Schrödinger to come up with his famous puzzle about the cat locked in a room (he used the German word for 'chamber', not 'box') with a diabolical device that is primed to kill the cat, but is in a 50:50 superposition of states. Updating his example, imagine that a detector in the room measures the spin of the electron. If it is up, the device is triggered and the cat dies. If it is down, the cat is safe. The electron is in a superposition of states before it is measured. But

* Quoted by Philip Ball.

47

there is nobody in the room to see what happens when the detector is triggered. So does the wave function collapse, or not? Is the cat also in a superposition of states, both dead and alive, until someone opens the door of the room to look in?

My own development of this idea involves two of the cat's off-spring (assuming it has survived) who I call Schrödinger's kittens.* These identical twin daughters of Schrödinger's cat live in identical space capsules, provided with all the necessities of life, and even some toys to play with. The capsules are connected by a tube, and in the middle of the tube there is a box which contains a single electron. The electron wave fills the box evenly. A partition is slid down to divide the box in two and separate the two capsules, each now connected to a box containing half an electron wave. The two capsules are now taken on separate long journeys, in opposite directions at exactly the same speed, until they are a couple of light years apart. Each one has a detector to monitor the presence of an electron. After a certain time (it doesn't have to be the same time in each case) the half-box in each capsule is opened by an automatic device. If there is an electron in it, the now grown-up cat dies. If not, the cat lives. But there is no intelligent observer to know what is going on. So are the cats now each in a superposition? An intelligent alien in a passing spaceship captures one of the capsules and looks inside, to see either a dead cat or a live cat. Is it at that point that the wave function in *each* capsule collapses, so that what the

---

* Particle physicists have taken the name and used it in another context. That is their privilege.

alien sees determines the fate of the other cat two light years away? Yes, according to the not so wonderful Copenhagen Interpretation.

So what is the alternative? There are many, although you may find them just as laughable as the CI, and the first off the rank is the one that started to emerge at the same time as the Copenhagen Interpretation, was nearly smothered at birth by Bohr, but lived to fight another day.

The Not So Impossible
Pilot Wave Interpretation

Louis de Broglie tried to resolve the puzzle of wave-particle dual-ity not by saying that an entity such as an electron could be either a wave or a particle, depending on how you looked at it, nor by saying that it was both wave and particle at the same time. He suggested that there might be two separate entities, a wave and a particle, which worked together to produce the effects we see in our experiments.

De Broglie was a pioneer of the idea of waves in quantum mechanics. It had been his suggestion that if, as Einstein had high-lighted, something previously thought of as waves (light) could also be regarded as particles (photons), then things previously regarded as particles (electrons) should also be treated as waves. This sug-gestion was soon confirmed by experiment, and led Schrödinger to develop his wave equation. It was natural that de Broglie should think deeply about the meaning of this wave-particle duality, and he put forward his solution to the puzzle at the same meeting in

Como where Bohr laid out the basics for what became known as the Copenhagen Interpretation.

In many ways, de Broglie's 'pilot wave' interpretation is the most natural and obvious way to explain wave-particle duality. He proposed that the wave and the particle are both real, and that the wave (which became known as a pilot wave) guides the particle to its destination, like a surfer riding waves in the sea. In the experiment with two holes, the pilot wave spreads out through both holes and interferes with itself to make a pattern of interfering waves. Particles that are fired through the experiment start out with slight differences in speed or direction, so they end up surfing in slightly different directions, following the waves to build up an interference pattern on the detector screen. We measure the properties of the particles, but we can never measure the properties of the wave, only infer its existence from the behaviour of the particles, which is hidden from us until they are detected. This kind of approach became known as 'hidden variables' theory.

A well-shuffled pack of cards provides a useful analogy. Imagine such a pack of cards small enough to be required to obey the rules of quantum physics, in a sub-microscopic device which enables you to turn the cards over one at a time to reveal their value. According to hidden variables theories, when you turn over the top card, the value you see is determined at random from the 52 possibilities allowed for by the pack. There is a 50:50 chance of getting a red card, a 1:52 chance of getting the five of clubs, and so on. The value of the card was hidden until you looked. But it always did have that value, even when you were not looking (in that sense, it is not really a variable!). After this first card has been seen – suppose it

Louis de Broglie
Getty Images

was indeed the five of clubs – there is now zero chance of finding the five of clubs, a 26:51 chance of finding a red card, and so on. Contrast this with the Copenhagen Interpretation, which says that the card does not have a value until you look. It is the act of looking which forces it to choose from the available possibilities. But in either case, if you keep turning cards over you will see the same sort of random pattern determined by the probabilities; you won't, for example, get the five of clubs twice. The experiment does not distinguish between the interpretations. But there is a huge difference in the explanation of what made that pattern.

David Lindley makes an analogy with a golfer practising on the putting green. He hits a series of golf balls all aimed at the same hole, but each one sets off at a slightly different speed and heading in a slightly different direction because of the inevitable minor variations in the golfer's putting technique. And the surface of the green is not perfectly smooth. So each ball follows a slightly different direction, and travels a slightly different distance. After the golfer has hit a hundred practice balls, they are spread out across the surface of the green in a pattern which has been determined by the irregularities of the surface that they have been travelling across. But the final position of each ball could be determined, in principle, if you knew the exact shape of the surface and the exact speed and direction the ball started with. In this sense, the Pilot Wave Interpretation is deterministic, and removes the element of chance associated with the collapse of the wave function, as well as removing the collapse of the wave function itself. Every particle has definite properties at all times. It's just that, like the cards in a well-shuffled pack, we don't know what those properties are until we look.

De Broglie spelled out the pilot wave argument in detail, not just the kind of vague discussion I have given, at the Como meeting. Looking back with the benefit of hindsight, in 1987 John Bell wrote in his book *Speakable and Unspeakable in Quantum Mechanics*: 'this idea seems so natural and simple, to resolve the wave-particle dilemma in such a clear and ordinary way, that it is a great mystery to me that it was so generally ignored.'

Actually, this is not such a great mystery. First, as I have mentioned, Bohr, aided and abetted by Wolfgang Pauli, poured scorn on the idea and crushed the more diffident de Broglie more by the force of their personalities and the force of their reputations than by the validity of their arguments. But reputation isn't everything. The second reason de Broglie's idea got trashed, along with other hidden variables theories, was von Neumann's incorrect 'proof' that such theories were impossible. De Broglie gave up any attempt to promote his idea, and it was so completely forgotten by physicists that when the American David Bohm came up with a similar idea at the beginning of the 1950s he didn't know anything about the earlier work. This initially led to some tension between him and de Broglie, who was annoyed not to be acknowledged, but this was smoothed over and the pilot wave idea is now often referred to as the de Broglie–Bohm interpretation.

The way Bohm got to his version of the pilot wave is particularly interesting in the present context. As a young researcher, Bohm wrote a textbook on quantum physics, published early in 1951, in which he spelled out the Copenhagen Interpretation to such good effect that even Pauli, a notoriously severe critic of anyone he regarded as his intellectual inferior (which meant everybody) approved of it. Einstein also felt that Bohm had done as good a job of explaining the CI as it was possible to do. But he got in touch with Bohm and stressed his own view that the CI was wrong. Bohm decided to see if there was another way of explaining what was going on in the quantum world, and soon found that there was. His pilot wave model was mathematically equivalent to the Copenhagen Interpretation, and gave the same answers to quantum questions as that interpretation; it was also essentially the same as de Broglie's model, but went slightly further in terms of describing the interaction between the quantum world and the classical world. But it was based on hidden variables, which von Neumann had said was impossible. Not least for that reason (but also, in the USA at least, because he was vilified as a communist sympathiser at the time of the McCarthy 'witch hunt'), Bohm was not taken seriously by many physicists, who thought that if von Neumann said it was impossible there must be a mistake in the model. There was one important exception.

In 1952 John Bell was working at the UK Atomic Energy Research Establishment at Malvern, in Worcestershire, and was chosen as one of the young scientists allowed to take a year off to do research. In his case, he went to work and study at Birmingham University, where he investigated quantum theory, and learned of Bohm's pilot wave idea. He immediately took the opposite view of most physicists. If Bohm's idea worked, and von Neumann said it was impossible, then it must mean that it was *von Neumann* who had made a mistake. Unfortunately, at that time von Neumann's book had only been published in German, which Bell did not read, and Bell had to get back to his day job designing particle accelerators,

David Bohm

Getty Images

before moving on to CERN in 1960. By 1963, von Neumann's book had been published in English, Bell found the mistake, and wrote up his findings during a sabbatical year in the United States. He also produced his own version of hidden variables theory as further proof that von Neumann was wrong. But as I have mentioned, he showed that all hidden variables theories, including the pilot wave idea, are non-local. As he wrote in one of the papers produced while in the USA, 'It is the requirement of locality, or more precisely that the result of a measurement on one system be unaffected by operations on a distant system with which it has interacted in the past, that creates the essential difficulty' with things like the EPR puzzle (or indeed, my kittens in space, where according to the de Broglie–Bohm theory the electron is always in one half-box and there is no superposition). In the Pilot Wave Interpretation, it is explicitly required that at any instant properties such as the velocity of one particle, or the way it changes the direction it is moving in, depend on the properties *at that same instant* of all the other particles it has interacted with.

Although I have not seen anyone else make the connection, this reminds me of a puzzle known as Mach's Principle. The physicist Ernst Mach – who was an influence on Einstein – drew attention to the puzzle, which had actually troubled scientists since at least Newton's day. It has to do with inertia. If you push something it resists being moved. I'm not talking about friction, but an idealised situation with an object floating freely in space. It will continue at rest or keep moving in a straight line (as Robert Hooke was the first to point out) until it is pushed, when it will change its speed, or its direction, or both. But how does it know that it is changing its direction or its speed? What is the change measured relative to? It does not take much observation to notice that inertia represents a resistance to change in motion relative to the Universe at large.

You don't have to imagine being in space to see the puzzle in all its glory. Isaac Newton himself, in his great book the *Principia*, described an actual experiment you can do in the privacy of your own home. He took a bucket of water hanging by its handle from a long rope, twisted it round and round and let go. The bucket started spinning, but at first the level of water in the bucket stayed the same. It didn't care that the bucket was moving relative to the water. Then, as the water picked up the rotation, it dipped in the centre, making a curved surface. When Newton grabbed the side of the bucket, it stopped spinning, but the water carried on spinning and the surface held its curved shape, gradually flattening as it slowed down. The shape of the surface of the water depended on how the water was moving relative to some mysterious fixed frame of reference, and had nothing to do with how it was moving relative to the bucket; this frame of reference is now identified as the average distribution of everything in the Universe. Actually, you don't even need a bucket to see the influence of the entire Universe on local things – just watch the surface of the liquid when you stir a cup of tea or coffee!

So the average distribution of everything in the Universe provides a frame of reference against which such changes are measured. Somehow, the 'local' object is influenced by everything 'out there'. Mach's Principle tells us that a particle's inertia is due to some interaction of that particle with all the other objects in the Universe. Just what that interaction is has long been a mystery. The Pilot

Wave Interpretation, and non-locality, may be the resolution of that puzzle.

This leads to an interesting conclusion, which also features in another interpretation (Solace 3). The de Broglie–Bohm Pilot Wave Interpretation applies to the whole Universe. The behaviour of a single particle here and now depends on the positions of every other particle in the Universe at this instant. But the implications are best explored in the context of that third Solace, the Many Worlds Interpretation. Before we move on, though, it's worth mentioning a surprising comment on Bohm's theory, from someone who might have been expected to endorse it. Even though Einstein had been the instigator of Bohm's attempt to find an alternative to the Copenhagen Interpretation, on 12 May 1952 he wrote to Max Born:

> Have you noticed that Bohm believes (as de Broglie did, by the way, 25 years ago) that he is able to interpret the quantum theory in deterministic form? That way seems too cheap to me.

Nobody is quite sure what he meant by this, but it highlights the confusion surrounding all of the interpretations of quantum mechanics.

The Excess Baggage
Many Worlds Interpretation

I f you have heard of the Many Worlds Interpretation (MWI), the chances are you think that it was invented by the American Hugh Everett in the mid-1950s. In a way that's true. He did come up with the idea all by himself. But he was unaware that essentially the same idea had occurred to Erwin Schrödinger half a decade earlier. Everett's version is more mathematical, Schrödinger's more philosophical, but the essential point is that both of them were motivated by a wish to get rid of the idea of the 'collapse of the wave function', and both of them succeeded.

As Schrödinger used to point out to anyone who would listen, there is nothing in the equations (including his famous wave equation) about collapse. That was something that Bohr bolted on to the theory to 'explain' why we only see one outcome of an experiment – a dead cat or a live cat – not a mixture, a superposition of states. But because we only *detect* one outcome – one solution to the wave function – that need not mean that the alternative solutions do not exist. In a paper he published in 1952, Schrödinger pointed out the ridiculousness of expecting a quantum superposition to collapse just because we look at it. It was, he wrote, 'patently absurd' that the wave function should 'be controlled in two entirely different ways, at times by the wave equation, but occasionally by direct interference of the observer, not controlled by the wave equation'.

Although Schrödinger himself did not apply his idea to the famous cat, it neatly resolves that puzzle. Updating his terminology, there are two parallel universes, or worlds, in one of which the cat lives and in one of which it dies. When the box is opened in one universe, a dead cat is revealed. In the other universe, there is a live cat. But there always were two worlds, which had been identical to one another until the moment when the diabolical device determined the fate of the cat(s). There is no collapse of the wave function. Schrödinger anticipated the reaction of his colleagues in a talk he gave in Dublin, where he was then based, in 1952. After stressing that when his eponymous equation seems to describe different possibilities they are 'not alternatives but all really happen simultaneously', he said:

Nearly every result [the quantum theorist] pronounces is about the probability of this or that or that ... happening – with usually a great many alternatives. The idea that they may not be alternatives but all really happen simultaneously seems lunatic to him, just impossible. He thinks that if the laws of nature took this form for, let me say, a quarter of an hour, we should find our surroundings rapidly turning into a quagmire, or sort of a featureless jelly or plasma, all contours becoming blurred, we ourselves probably becoming jelly fish. It is strange that he should believe this. For I understand he grants that unobserved nature does behave this way – namely according to the wave equation. The aforesaid alternatives come into play only when we make an observation – which need, of course, not be a scientific observation. Still it would seem that, according to the quantum theorist, nature is prevented from rapid jellification only by our perceiving or observing it … it is a strange decision.

In fact, nobody responded to Schrödinger's idea. It was ignored and forgotten, regarded as impossible. So Everett developed his own version of the MWI entirely independently, only for it to be almost as completely ignored. But it was Everett who introduced the idea of the Universe 'splitting' into different versions of itself when faced with quantum choices, muddying the waters for decades.

Everett came up with the idea in 1955, when he was a PhD student at Princeton. In the original version of his idea, developed in a draft of his thesis which was not published at the time, he compared the situation with an amoeba that splits into two daughter cells. If amoebas had brains, each daughter would remember an identical history up until the point of splitting, then have its own personal memories. In the familiar cat analogy, we have one universe, and one cat, before the diabolical device is triggered, then two universes, each with its own cat, and so on. Everett's PhD supervisor, John Wheeler, encouraged him to develop a mathematical description of his idea for his thesis, and for a paper published in the *Reviews of Modern Physics* in 1957, but along the way the amoeba analogy was dropped and did not appear in print until later. But Everett did point out that since no observer would ever be aware of the existence of the other worlds, to claim that they cannot be there because we cannot see them is no more valid than claiming that the Earth cannot be orbiting around the Sun because we cannot feel the movement.

Everett himself never promoted the idea of the MWI. Even before he completed his PhD, he had accepted the offer of a job at the Pentagon working in the Weapons Systems Evaluation Group on the application of mathematical techniques (the innocently titled game theory) to secret Cold War problems (some of his work was so secret that it is still classified) and essentially disappeared from the academic radar. It wasn't until the late 1960s that the idea gained some momentum, when it was taken up and enthusiastically promoted by Bryce DeWitt, of the University of North Carolina, who wrote: 'every quantum transition taking place in every star, in every galaxy, in every remote corner of the universe is splitting our local world on Earth into myriad copies of itself.' This became too much for Wheeler, who backtracked from his original endorsement of the MWI and in the 1970s said: 'I have reluctantly had to give up my support of that point of view in the end – because I am afraid it carries too great a load of metaphysical baggage.'* Ironically, just at that moment the idea was being revived and transformed, through applications in cosmology and quantum computing.

The power of the interpretation began to be appreciated even by people reluctant to endorse it fully. John Bell noted that 'persons of course multiply with the world, and those in any particular branch

---

* Quoted in *Some Strangeness in the Proportion*, ed. H. Woolf (Addison-Wesley, 1981).

would experience only what happens in that branch', and grudgingly admitted that there might be something in it:

> The 'many worlds interpretation' seems to me an extravagant, and above all an extravagantly vague, hypothesis. I could almost dismiss it as silly. And yet ... It may have something distinctive to say in connection with the 'Einstein Podolsky Rosen puzzle', and it would be worthwhile, I think, to formulate some precise version of it to see if this is really so. And the existence of all possible worlds may make us more comfortable about the existence of our own world ... which seems to be in some ways a highly improbable one.*

The precise version of the MWI came from David Deutsch, in Oxford, and in effect put Schrödinger's version of the idea on a secure footing, although when he formulated his interpretation Deutsch was unaware of Schrödinger's version. Deutsch worked with DeWitt in the 1970s, and in 1977 he met Everett at a conference organised by DeWitt – the only time Everett ever presented his ideas to a large audience. Convinced that the MWI was the right way to understand the quantum world, Deutsch became a pioneer in the field of quantum computing, not through any interest in computers as such, but because of his belief that the existence of a working quantum computer would prove the reality of the MWI.

---

* *Speakable and Unspeakable in Quantum Mechanics* (Cambridge University Press, 1987).

This is where we get back to a version of Schrödinger's idea. In the Everett version of the cat puzzle, there is a single cat up to the point where the device is triggered. Then the entire Universe splits in two. Similarly, as DeWitt pointed out, an electron in a distant galaxy confronted with a choice of two (or more) quantum paths causes the entire Universe, including ourselves, to split. In the Deutsch–Schrödinger version, there is an infinite variety of universes (a Multiverse) corresponding to all possible solutions to the quantum wave function. As far as the cat experiment is concerned, there are many identical universes in which identical experimenters construct identical diabolical devices. These universes are identical up to the point where the device is triggered. Then, in some universes the cat dies, in some it lives, and the subsequent histories are correspondingly different. But the parallel worlds can never communicate with one another. Or can they?

Deutsch argues that when two or more previously identical universes are forced by quantum processes to become distinct, as in the experiment with two holes, there is a temporary interference between the universes, which becomes suppressed as they evolve. It is this interaction which causes the observed results of those experiments. His dream is to see the construction of an intelligent quantum machine – a computer – which would monitor some quantum phenomenon involving interference going on within its 'brain'. Using a rather subtle argument, Deutsch claims that an intelligent quantum computer would be able to remember the experience of temporarily existing in parallel realities. This is far from being a practical experiment. But Deutsch also has a much simpler 'proof' of the existence of the Multiverse.

David Deutsch
Getty Images

What makes a quantum computer qualitatively different from a conventional computer is that the 'switches' inside it exist in a superposition of states. A conventional computer is built up from a collection of switches (units in electrical circuits) which can be either on or off, corresponding to the digits 1 or 0. This makes it possible to carry out calculations by manipulating strings of numbers in binary code. Each switch is known as a bit, and the more bits there are, the more powerful the computer is. Eight bits make a byte, and computer memory today is measured in terms of billions of bytes – gigabytes, or Gb. Strictly speaking, since we are dealing in binary, a gigabyte is 2^{30} bytes, but that is usually taken as read. Each switch in a quantum computer, however, is an entity which can be in a superposition of states. These are usually atoms, but you can think of them as being electrons that are either spin up or spin down. The difference is that in the superposition, they are both spin up and spin down at the same time – 0 *and* 1. Each switch is called a qubit, pronounced 'cubit'.

Because of this quantum property, each qubit is equivalent to two bits. This doesn't look impressive at first sight, but it is. If you have three qubits, for example, they can be arranged in eight ways: 000, 001, 010, 011, 100, 101, 110, 111. The superposition embraces all these possibilities. So three qubits are not equivalent to six bits (2×3), but to eight bits (2 raised to the power of 3). The equivalent number of bits is always 2 raised to the power of the number of qubits. Just 10 qubits would be equivalent to 2^{10} bits, actually 1,024 but usually referred to as a kilobit. Exponentials like this rapidly run away with themselves. A computer with just 300 qubits would be equivalent to a conventional computer with more bits than there are atoms in the observable Universe. How could such a computer carry out calculations? The question is more pressing since simple quantum computers, incorporating a few qubits, have already been constructed and shown to work as expected. They really are more powerful than conventional computers with the same number of bits.

Deutsch's answer is that the calculation is carried out simultaneously on identical computers in each of the parallel universes corresponding to the superpositions. For a three-qubit computer, that means eight superpositions of computer scientists working on the same problem using identical computers to get an answer. It is no surprise that they should 'collaborate' in this way, since the experimenters are identical, with identical reasons for tackling the same problem. That isn't too difficult to visualise. But when we build a 300-qubit machine – which will surely happen – we will, if Deutsch is right, be involving a 'collaboration' between more universes than there are atoms in our visible Universe. It is a matter of choice whether you think that is too great a load of metaphysical baggage. But if you do, you will need some other way to explain why quantum computers work.

Most quantum computer scientists prefer not to think about these implications. But there is one group of scientists who are used to thinking of even more than six impossible things before breakfast – the cosmologists. Some of them have espoused the Many Worlds Interpretation as the best way to explain the existence of the Universe itself.

Their jumping-off point is the fact, noted by Schrödinger, that there is nothing in the equations referring to a collapse of the wave function. And they do mean *the* wave function; just one, which describes the entire world as a superposition of states – a Multiverse made up of a superposition of universes.

The first version of Everett's PhD thesis (later modified and shortened on the advice of Wheeler) was actually titled 'The Theory of the Universal Wave Function'.* And by 'universal' he meant literally that, saying:

> Since the universal validity of the state function description is asserted, one can regard the state functions themselves as the fundamental entities, and one can even consider the state function of the whole universe. In this sense this theory can be called the theory of the 'universal wave function,' since all of physics is presumed to follow from this function alone.

... where for the present purpose 'state function' is another name for 'wave function'. 'All of physics' means everything, including us – the 'observers' in physics jargon. Cosmologists are excited by this not because they are included in the wave function, but because this idea of a single, uncollapsed wave function is the only way in which the entire Universe can be described in quantum mechanical terms while still being compatible with the general theory of relativity. In the short version of his thesis published in 1957, Everett concluded that his formulation of quantum mechanics 'may therefore prove a fruitful framework for the quantization of general relativity'.

....

* It was eventually published in 1973 in the volume *The Many-Worlds Interpretation of Quantum Mechanics*, edited by B.S. DeWitt and N. Graham (Princeton University Press).

Although that dream has not yet been fulfilled, it has encouraged a great deal of work by cosmologists since the mid-1980s, when they latched on to the idea. But it does bring with it a lot of baggage.

The universal wave function describes the position of every particle in the Universe at a particular moment in time. But it also describes every possible location of those particles at that instant. And it also describes every possible location of every particle at any other instant of time, although the number of possibilities is restricted by the quantum graininess of space and time. Out of this myriad of possible universes, there will be many versions in which stable stars and planets, and people to live on those planets, cannot exist. But there will be at least some universes resembling our own, more or less accurately, in the way often portrayed in science fiction stories. Or, indeed, in other fiction. Deutsch has pointed out that according to the MWI any world described in a work of fiction, provided it obeys the laws of physics, really does exist somewhere in the Multiverse. There really is, for example, a *Wuthering Heights* world (but not a *Harry Potter* world).

That isn't the end of it. The single wave function describes all possible universes at all possible times. But it doesn't say anything about changing from one state to another. Time does not flow. Sticking close to home, Everett's parameter, called a state vector, includes a description of a world in which we exist, and all the records of that world's history, from our memories, to fossils, to light reaching us from distant galaxies, exist. There will also be another universe exactly the same except that the 'time step' has been advanced by, say, one second (or one hour, or one year). But there is no suggestion that any universe moves along from one time step to another. There will be a 'me' in this second universe, described by the universal wave function, who has all the memories I have at the first instant, plus those corresponding to a further second (or hour, or year, or whatever). But it is impossible to say that these versions of 'me' are the same person. Different time states can be ordered in terms of the events they describe, defining the difference between past and future, but they do not change from one state to another. All the states just exist. Time, in the way we are used to thinking of it, does not 'flow' in Everett's MWI.

As far as my perception is concerned, though, it is time for a change. Time to seek another kind of solace, this time in decoherence.

The Incoherent
Decoherence Interpretation

In order to have decoherence, something must be coherent in the first place. Physicists have a clear understanding of what they mean by coherence, and proponents of the Decoherence Interpretation argue that it is, in fact, coherence that makes the quantum world behave the way it does.

As usual, the experiment with two holes highlights what is going on. The light (or whatever) waves that are spreading out from the two holes originally came from a single source, and are therefore in step with one another. The holes simply send the coherent waves along different paths, and differences in the lengths of those paths affect the way the two sets of waves interact with one another – in step *here*, out of step *there*. There is a regular pattern in the ups and downs of the waves, which makes it possible for them to interfere with one another in such a way that they will produce a regular pattern of light and shade. If there is no coherence between the waves, as with light spreading out from two torches and shining directly on to a wall, there is no interference pattern – there is interference, but everything is so jumbled up that there is no pattern. According to the Decoherence Interpretation, it is when things get jumbled up that 'quantumness' disappears. But the light from the two torches never was coherent in the first place. It was incoherent. There's another useful analogy – the 'Mexican wave' you sometimes see in sports arenas. If everybody in the arena raises their arms at random, all you see is a jumble of waving hands. But if each person raises and lowers their arms at the right time, copying their neighbours, a wave sweeps round the stadium. The wave is coherent; the random waving is incoherent. So the term 'decoherence' is not all that apt in the quantum context. It might make more sense to call this model the incoherent interpretation of quantum mechanics; but its aficionados might feel that this would give the wrong impression of their favoured idea!

If those aficionados are right, the boundary between quantumness and the everyday world depends on coherence, not on size. Bohr and his colleagues were necessarily vague about this. They could argue, quite reasonably, that an object as large and complicated as a cat was too big to be in a quantum superposition, even though individual atoms could be in a superposition. But in imagining variations on the cat in the box thought experiment, where did you draw the line? Was a flea big enough to know if it was dead or alive, or in a superposition? A microbe? Nobody knew.

One man took up the challenge of finding out. Anthony Leggett, who worked at the University of Sussex in the late 1960s and 1970s, determined to devise experiments to test whether the rules of quantum mechanics could still be used to describe the behaviour of so-called 'macroscopic' objects, things big enough to hold in your hand, or bigger. This led him to develop devices called SQUIDs (from Superconducting Quantum Interference Device). The archetypal SQUID was about the size of a wedding ring, so you could indeed hold it in your hand,* but it had to be supercooled to operate, so you couldn't hold it while it was working. An electric current circulating in a superconductor flows along for ever, once it gets going, and the behaviour of such a current circulating round a SQUID can be monitored and tweaked using electric and magnetic fields. These experiments show that the electron wave going round the ring behaves like a single quantum entity, about a hundred million times bigger than an atom (and certainly much bigger than a bacterium, or even a flea). Leggett had achieved his first objective. But there is more. You might think that the wave could flow one way round the ring, or the other way round the ring, but not both ways at the same time. You would be wrong. Experiments carried out early in the 21st century showed the effects of the wave going both ways round the ring at the same time. Not two different waves going in opposite directions, but the *same* wave going both ways at once – a superposition. It is not the size of the object that determines the object's quantumness, but the fact that the waves are coherent.

This work has progressed enormously since the early days, along the way earning Leggett both a Nobel Prize and a knighthood; SQUIDs are being scaled up in size, have practical applications in

* I have.

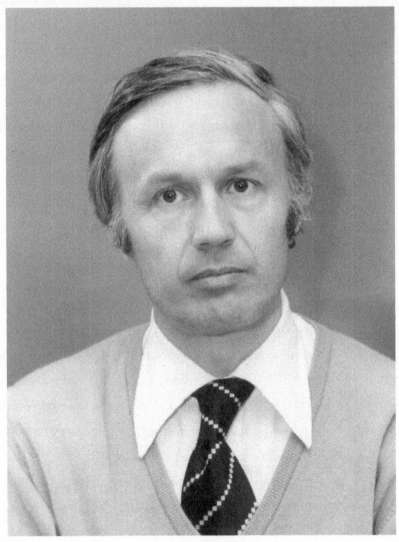

Anthony Leggett
Science Photo Library

medicine as sensitive detectors of magnetic fields produced by the human body, and are potential components of quantum computers. For now, though, what matters is that they behave as a macroscopic example of distinct quantum states when the waves are coherent, but cease to exhibit quantumness when they warm up and the waves decohere. In the language of Bohr, it seems that decoherence causes the 'collapse of the wave function'. Some people assume that this simply means that the Decoherence Interpretation is only the Copenhagen Interpretation by another name. But this assumption ignores the key role of superposition and entanglement in the strict Decoherence Interpretation.

Superposition and entanglement are two sides of the same coin. When two 'particles' interact with one another they become entangled, and for ever afterwards what happens to one of the particles affects the other. In effect, they are now a single entity. In a similar way, the single wave going both ways at once around a SQUID ring can be thought of as two waves in a superposition, entangled with one another. The result is a single quantum entity, a wave that goes not one way but two. It is no surprise that the Decoherence Interpretation only emerged around the same time as the experiments in the 1980s which established that entanglement is a valid description of the way the world works.

So what actually happens when a 'pure' quantum entity interacts with the outside world and 'decoheres'? It doesn't get *less* entangled, but *more*. Imagine a lonely particle in a pure quantum state. As soon as it bounces off another particle (or even if it interacts with a photon of light) it becomes entangled. If either of the two entangled entities interact with a third entity, all three become entangled, with their quantum states in a superposition. The entanglement spreads faster than the proverbial forest fire, so that in practice there is no such thing as a 'pure' quantum system separated from the world outside (except in very special circumstances like the SQUID experiments), but an entangled system of both, a superposition of everything that has interacted with the original particle and everything it has ever interacted with, and everything that that everything has ever interacted with or has ever come into contact with. 'Decoherence' actually involves linking everything in the entire world – the Universe – into a single quantum system. We no longer detect the quantumness of the once-isolated particle because it is mixed up with everything else. The resulting incoherence makes it extremely difficult to unravel the underlying quantumness of anything except the simplest systems. The mathematicians will tell you that this might be possible in principle, since the equations that describe the quantum world are time-reversible. But don't hold your breath waiting for someone to do the experiment.

As Philip Ball has pointed out, decoherence very quickly produces an incoherent state equivalent to a superposition of more quantum states than there are fundamental particles in the observable Universe. He asks: 'can a problem be said to be strictly impossible solely because there is not enough information available in the universe to solve it?'* Ball also gives some estimates of how long it takes for a system to decohere. Decoherence happens more quickly for

---

* David Deutsch, of course, would not see this as a problem at all! But this Solace is primarily concerned with decoherence, not the MWI.

larger objects, because there are more bits in them to interact with other things, and with each other. For a dust grain floating in the air and being buffeted by the molecules around it, decoherence takes less time than that required for a photon, travelling at the speed of light, to cross a distance equivalent to the diameter of a proton. Even in interstellar space, floating freely and interacting with nothing except the photons of the cosmic microwave background radiation, the dust grain will decohere in about a second. 'Decoherence is to all practical purposes instant and inevitable.' And this applies to Schrödinger's famous cat. In order to be 'both dead and alive', the cat would have to be 'prepared' in some incredibly unlikely coherent state of pure quantumness. It is one thing to prepare a SQUID in a pure quantum state, but quite another to do that to a cat. And if you did, it would decohere into either a dead cat or a living cat faster than a dust grain floating in the air decoheres.

This also pulls the rug from under one of the philosophical objections to the Copenhagen Interpretation. Taken at face value, the CI says that 'nothing is real' unless it is being observed or measured. Things like cats in boxes can exist in superpositions of states. So, said opponents of the idea, does the Moon exist when nobody is looking at it, or is it in a superposition of all possible quantum states? Did it exist, in this sense, before there was life on Earth? Bohr had no satisfactory answer to this. The Decoherence Interpretation has – photons from the microwave background radiation, let alone those of sunlight, are more than adequate to produce decoherence and make the Moon 'real'.

This, though, is not the end of the decoherence story. Instead of applying the idea only to the here and now (whatever 'here and now' means in an entangled Universe), some people have applied this way of thinking to the entire history – or histories – of the Universe. What used to be a separate interpretation in its own right, the Consistent Histories Interpretation, is now the Decoherent Histories Interpretation. But I will start with the 'consistency' bit of the story.

This harks back to the idea that the only things we know about the quantum world (or the world at large) are what we can see and measure. In advance of carrying out an experiment, or making a measurement, we can only calculate the probability of different outcomes of the experiment. But once we have made the measurement, we have a definite result, in some sense selected from the array of possibilities. The argument of the Consistent Histories approach is that whatever the result of that measurement – whatever *anything* that happens in the world – it must be consistent with the past, with history. So when we look at the interference pattern produced in the experiment with two holes, all we can say is that the pattern is *consistent with* waves having gone through the holes and interfered with one another. When an electron is knocked out of a metal surface by light, we can only say that this is *consistent with* the light arriving in the form of a photon.

The cosmological implications of all this have been widely discussed, notably by Stephen Hawking and his colleagues. Hawking described the traditional way of trying to understand the origin of the Universe in quantum terms as the 'bottom up' approach. You start by guessing what the Universe could have looked like in the beginning, as a superposition of wave functions, and try to work out how it got from that state to the state we see it in today. He preferred the alternative 'top down' approach, in which you start out from where we are now and work backwards, in a consistent way, to determine which wave functions contributed to its origin.

The trouble is, there may be (usually will be) more than one different way to arrive at the result we observe – more than one consistent history. There is no unique 'history of the Universe' revealed by this approach. If, in the electron version of the double-slit experiment, an electron arrives at a definite point on the detector screen, there is no way to tell which hole it has passed through. Both histories are consistent with what we observe. And the world at large is much more complicated than the experiment with two holes, allowing for a much wider choice of consistent histories. I'll get back to that. But first, where does decoherence come into the story?

If every 'measurement' – every quantum interaction – selects from an array of possible histories, then we can imagine working backwards in time, all the way to the Big Bang (and maybe further, but I won't go into that here), with decoherence (for that is what it is) sifting out versions of history that are consistent. In the beginning, anything is possible. But as soon as any quantum interaction occurs, some possibilities are ruled out, and the variety of different universes is reduced. That is, the range of consistent *past* universes is reduced. This continues all the way up to the present, selecting the history of our universe (but crucially not only our universe) from the worlds of possibility. The Decoherent Histories approach does not select a unique Universe. We are back to a variation on the theme of Many Worlds, arrived at by a different route.

The possibility of using decoherence to transform the Many Worlds idea into a 'Many Histories' idea seemed to some physicists to remove the baggage of many parallel worlds, all equally real, and replace it with different histories, which only existed as ghost-like states among the probabilities. But in the mid-1990s it became apparent that things were not so simple. After attending a conference where he heard Fay Dowker discuss the possibilities, Lee Smolin, who works at the Perimeter Institute in Canada, experienced a flash of insight which he later described in his book *Three Roads to Quantum Gravity:**

> While the 'classical' world we observe, in which particles have definite positions, may be one of the consistent worlds described by a solution to the theory, Dowker and [Adrian] Kent's results showed that there had to be an infinite number of other worlds, too. Moreover, there were an infinite number of consistent worlds that have been classical up to this point but will not be anything like our world in five minutes' time. Even more disturbing, there were worlds that were classical now but were arbitrarily mixed up superpositions of classical [worlds] at any point in the past ... if the consistent histories interpretation is correct, we have no right to deduce from the existence of fossils now that dinosaurs roamed the planet a hundred million years ago.

All of the histories are equally real, and what we perceive as 'the' history of our world depends on the questions we ask of the world. In just the same way that when we experiment with electrons, if we look for waves we get waves, but if we look for particles we get

....

* (Weidenfeld & Nicolson, London, 2000). Hard stuff.

particles, if we look for evidence of the existence of dinosaurs in the past then we find a history consistent with the existence of dinosaurs in the past. This does not necessarily mean that there 'really were' dinosaurs in the past; only that the state of the world today is consistent with the possibility that there were dinosaurs in the past. As Smolin has put it, we have 'a theory in which we can formulate the answers, but not the questions'.

Not quite all things to all people, but, depending on taste, you can see the Decoherence Interpretation either as a version of the Copenhagen Interpretation, or as a version of the Many Worlds Interpretation. But don't worry if none of this is to your taste. Perhaps you can find solace in the Ensemble Interpretation.

SOLACE

The Ensemble Non-Interpretation

The Ensemble Interpretation was the first, and simplest, alternative to the Copenhagen Interpretation, and the one favoured by Albert Einstein, who said:

> The attempt to conceive the quantum-theoretical description as the complete description of the individual systems leads to unnatural theoretical interpretations, which immediately become unnecessary if one accepts the interpretation that the description refers to ensembles of systems and not to individual systems.*

Leslie Ballentine, a leading modern proponent of the idea who works at Simon Fraser University in Canada, explains that Einstein's 'criticism of the interpretation accepted, at least tacitly, by many physicists was that the quantum state [wave] function does not provide a description of an individual system but rather of an ensemble

---

* See *Albert Einstein: Philosopher-Scientist*, ed. P.A. Schilpp (Harper & Row, New York, 1949).

of similar systems'. But this 'interpretation' doesn't really interpret anything at all. It simply says that everything that looks odd about the quantum world can be explained in terms of statistics (it is sometimes known as the statistical interpretation). It is more like the policeman at a crime scene who instructs the crowd of onlookers, 'Nothing for you to see here, move along, please.'

The statistics are those of the ensembles. But the ensembles are not the kind that spring to the mind of most people when they hear the term. In everyday language, an ensemble is a group of things that have some common property, or are working together – such as a musical string ensemble. To a statistician, a collection of 600 identical dice could constitute an ensemble, and if all those dice were rolled together then the laws of probability would lead us to expect to see near enough 100 sixes, 100 fives, 100 fours, 100 threes, 100 twos and 100 ones. But there is another way to get the same statistical outcome. Take a single perfect die, and roll it 600 times. You would expect 6 to come up about a hundred times, 5 to come up about a hundred times, and so on. This is the kind of ensemble the quantum physicists are referring to. A box full of molecules of gas would not constitute an ensemble in this sense; but many identical boxes of gas each experimented on in the same way would. Ideally, you would carry out exactly the same experiment on exactly the same particle many times, and monitor the outcome of each of these 'trials'. *That* is the ensemble. The results would follow a probability distribution in accordance with the rules developed by Max Born.

It would be very hard to carry out such an idealised experiment, but that isn't really the point. Instead of, say, a million electrons going through the double-slit experiment at the same time and being detected on the other side, think of the same electron going round and round a million times, with the position it arrives at on the other side being noted each time it goes past. The crucial point which proponents of this interpretation like is that the particles are always real particles in the everyday use of the term. The wave function does not apply to individual particles, so that each individual electron, for example, really is either spin up or spin down, but when you have many particles the probability of finding either possibility when you examine an individual electron is (other things being equal) 50:50. There is no wave-particle duality, no superposition, and no dead-and-alive cats. It would be hard to carry out the cat experiment a hundred times or more using the same cat, of course, but if you did it with a hundred cats one after the other, according to the Ensemble Interpretation half of them would live, and half of them would die, but none of them would be in a superposition.

It sounds tempting. Common sense. But as Euan Squires has pointed out, we must not 'claim that we have solved the problems [of interpretation]. We have merely ignored them ... individual systems exist.' And how is it supposed to work in practice? As is often the case in quantum theory, the waters become muddier once you try to work out what happens when the system – in this case the ensemble – is studied, or otherwise interacts with the outside world. Preparing the system involves a certain amount of randomness, and observing it involves another layer of randomness. We are back with the problem of where the system ends and the outside world begins, like the entanglement that spreads out across the

Universe in the Decoherence Interpretation. An example of this interaction with the outside world that is sometimes put forward in support of the Ensemble Interpretation is the so-called 'watched pot' experiment.

The key to this idea is that although the equations of quantum physics describe the probability of finding a system in one state or another, they say nothing about systems making a transition *from* one state to another. There is nothing in the equations describing the 'collapse of the wave function'. And no experiment has ever caught a wave function in the act of collapse. Back in 1954, Alan Turing pointed out that a quantum system that is constantly 'watched' will never change. He wrote:

> It is easy to show using standard theory that if a system starts in an eigenstate* of some observable, and measurements are made of that observable N times a second, then, even if the state is not a stationary one, the probability that the system will be in the same state after, say, one second, tends to one as N tends to infinity; that is, that continual observations will prevent motion.[†]

Physicists try to explain this in various ways. Here's one. Imagine a system in a well-defined state with a wave of probability spreading out and gradually increasing the probability of finding it in some other state. If you wait a long time, then look, you probably observe it in a different state. But if you look very quickly, the probability

....

* A quantum-mechanical state corresponding to a single value of a wave equation.
[†] Quoted by Andrew Hodges in *Alan Turing: Life and Legacy of a Great Thinker* (Hutchinson, London, 1983).

has not had time to change, and then it will still be in the same state. It cannot be in an intermediate state, because there are no intermediate states. So the wave has to start spreading out again from the same position. Look frequently enough, and it will never get anywhere. The quantum 'pot' will not boil if you keep looking at it. That was Turing's prediction, and it has now been tested by experiments.

These experiments involve variations on a theme. Typically, the 'pot' is a few thousand ions of an element such as beryllium, trapped by electric and magnetic fields. An ion is an atom from which one or more electrons have been stripped, leaving it with a positive charge which makes it easy to manipulate with such fields. The ions can be prepared in an energy state from which they 'want' to escape, jumping down to a lower energy state. The state of the system can be monitored by a subtle technique involving lasers, to find out how many ions have decayed in this way after a certain time.

In one typical experiment, after 128 milliseconds half the ions had decayed. But if the laser 'looked' after just 64 milliseconds, only a quarter of the ions had gone. If the laser flickered once every 4 milliseconds, looking 64 times in 256 milliseconds, almost all the ions were still in their original state. In terms of the probabilities corresponding to the wave function, this failure to 'boil' is because after 4 milliseconds the probability of an ion making the transition was only 0.001 per cent, so 99.99 per cent of the ions had to still be in Level 1. And this holds for every 4 millisecond interval. The shorter the time interval between observations, the stronger the effect. Wave functions never collapse when they are watched. So why expect them to collapse at all? Ballentine argues that they don't, and that this is experimental evidence in support of the Ensemble Interpretation.

There is, though, one big problem with the Ensemble Interpretation. It specifically says that the wave function does not apply to individual quantum entities, and that there is no such thing as a superposition of states. But experimenters now routinely manipulate individual quantum entities, such as electrons, in situations (such as quantum computing) where they seem to be following the wave function description, and a SQUID ring seems able to demonstrate a macroscopic single quantum entity (the electron wave going both ways at once) which is in a superposition. I used to think that this was a death blow to the idea. But Lee Smolin has revived it in a new incarnation.

This new version of the Ensemble Interpretation fully embraces the concept of non-locality, which is now shown by experiment to be a key feature of the Universe. Einstein probably would not be happy with this metamorphosis of the interpretation he endorsed. But Smolin is so happy with it that, with characteristic chutzpah, he calls it the Real Ensemble Interpretation. The key difference is that while in the traditional Ensemble Interpretation the members of the ensemble do not actually all exist at the same time, in Smolin's version they are all simultaneously real. There's a piece of jargon here which needs to be got out of the way to make this point more succinctly. The possible quantum components of the ensemble (hydrogen atoms, say) are called 'beables' because they are things that could possibly be. But as with the case of rolling a single die 600 times rather than rolling 600 dice all at once, they are not all in existence together. What out of deference to Smolin I shall refer

Lee Smolin
Nir Bareket

to as the REI says is that the beables that make an ensemble really are in being simultaneously, like the 600 dice rolled together, not like the same die rolled 600 times. There is a real state of affairs in any quantum system at any given time, determined by the values of the beables.

Smolin starts from the reasonable principle that anything that is supposed to influence the behaviour of a real system in the Universe must itself be a real system in the Universe. It is not acceptable, he says, 'to imagine that there is a spooky way in which "potentialities affect realities"'. In the Pilot Wave Interpretation, for example, the wave is a real feature of the Universe, a beable, not some spooky 'wave of probability'. But that interpretation runs foul of another postulate put forward by Smolin, that nowhere in nature should there be an 'unreciprocated action'. This is an extension of Newton's law that in classical systems action and reaction are equal and opposite. In the Pilot Wave Interpretation, the wave influences the particle, but the particle does not influence the wave – it does not reciprocate. But in the ensembles pictured by Smolin, all the beables of an ensemble influence each other reciprocally, to produce the behaviour we see in experiments like the one with two holes. And if all the components of the ensemble are real, there is no reason why there cannot be new (as in the sense of previously undiscovered) interactions between them.

He gives an example, involving hydrogen atoms in their lowest energy state, called the ground state. There is an ensemble of every such hydrogen atom in the Universe – a real ensemble of real beables. These components of the ensemble interact with one another in a non-local way, in which the beables copy each others' states in accordance with the rules of probability associated with those quantum states. The probabilities for the copying processes do not depend on where the components are in space, but they do depend on the way the beables are distributed in the ensemble. So the quantum statistics make it possible to have a list of positions where hydrogen atoms in their ground states will be found, but they do not tell us which hydrogen atom is in which location. Smolin has been able to show mathematically that with a few such simple rules about how pairs of beables influence each other, this process can produce all of the observed behaviour of quantum systems. And it can also explain why things like cats and people cannot be in a superposition.

Quantum mechanics, says Smolin, applies to small subsystems of the Universe which come in many copies, like hydrogen atoms in their ground state. But macroscopic systems like cats and people have no copies anywhere in the Universe, so they are not affected by the copying process that involves interacting quantum beables. They have nothing to interact with, in this sense.

This has some interesting implications. First, the Universe must be finite. In an infinite universe, there would be infinitely many copies of you, so the interactions described by Smolin's equations would affect you, and you would behave like a quantum particle! Secondly, as well as deriving the Schrödinger wave equation from his simple mathematical rules, Smolin can also derive the laws of classical mechanics – Newton's laws, and so on – as an approximation to quantum mechanics. But he suspects that quantum mechanics is itself an approximate version of some deeper description of the Universe (indeed, this was his real motive for delving into these murky waters), and he goes so far as to suggest that genuine faster-than-light signalling might occur if that is the case. A strong hint that we do not yet have the ultimate theory is that, as you may have noticed, the interaction between beables seems to imply that there is a unique cosmic time, so that the interactions can occur simultaneously, which would require an extension to relativity theory.* 'Quantum physics', he says, 'must be an approximation to a cosmological theory which is formulated in different terms.' The place to look for those underlying laws might be in experiments involving systems that are likely to exist in small numbers of copies in the Universe, in the borderland between the microscopic and the macroscopic worlds. Experiments with things like quantum computers might make it possible to tell whether there are any copies of them in the Universe. There might be real observable effects arising from corrections to quantum physics that depend on the size of the ensemble.

If all this sounds bizarre, Smolin has a reminder for us. At one time, people found it impossible to believe that the Sun influenced the dynamical behaviour of the planets, because that would involve a strange action at a distance. As I mentioned earlier, even Newton didn't try to explain how it worked, famously declaring 'Hypotheses non fingo' – I do not make hypotheses. REI involves a 'new' kind of non-local interaction between beables, but this should be no more alarming than the fact that just over a hundred years ago the explanation of the interaction between the Sun and Earth involved

* FYI, there are theorists who think that there is a preferred measure of simultaneity in the general theory of relativity, but those waters are too deep for me to plunge into here.

a 'new' kind of interaction, now described by the general theory of relativity. Non-locality seems spooky to non-physicists because they are not used to it, but to a growing number of physicists it is now as much an accepted fact as the fact of gravity. Not much to digest before breakfast after all. Interactions that ignore space are an established feature of the world. But what about interactions that ignore time? Can we seek solace there?

The Timeless Transactional Interpretation

The Transactional Interpretation of quantum mechanics (TI) has its roots in a puzzle about the nature of light which intrigued Albert Einstein. Since it was puzzling over the nature of light that led Einstein to develop the special theory of relativity, this alone makes it worth taking seriously. The realisation that led to the special theory is that the equations which describe the behaviour of light, and all other electromagnetic radiation, say that the speed of light is the same for everyone, a constant now written as c. If you shine a torch at me and I am standing next to you, I will measure the speed of the light from the torch as c. But even if I am whizzing towards you, or away from you, at high speed, I still measure the speed of the light from the torch as c. From this simple fact Einstein developed relativity theory.

The equations which say, among other things, that the speed of light is the same for any observer are known as Maxwell's equations, after the nineteenth-century physicist who discovered them. But James Clerk Maxwell's equations have another curious property.

They are symmetrical in time. In any problem involving electro-magnetic radiation, such as the radiation associated with a moving electron, there are always two solutions to the equations. One describes a so-called 'retarded' wave, moving out from a source and forwards in time, to be absorbed somewhere out in the world. The other describes an 'advanced' wave, which comes from the absorb-ers out in the world and converges on what we think of as the source (in this case, the moving electron) from the future. Most physicists simply ignore this 'advanced solution'. But in 1909 Einstein said:

> In the first case the electric field is calculated from the totality of the processes producing it, and in the second case from the total-ity of processes absorbing it ... both kinds of representation can always be used, regardless of how distant the absorbing bodies are imagined to be. Thus one cannot conclude that the [retarded solution] is any more special than the solution [containing equal parts advanced and retarded waves].*

Regardless of how distant the absorbing bodies are imagined to be. This is not something that applies only to electrons interacting with their neighbours, but, for example, to the TV signals spreading out from Earth across the Universe. The equations that describe this process always include a solution describing advanced waves con-verging from the Universe onto the TV antennas the signals were broadcast from. There is a hint here of another (or the same?) kind

* See *The Collected Papers of Albert Einstein*, Volume 2, ed. A. Beck and P. Havas (Princeton University Press, 1989); also quoted in John Cramer's *The Quantum Handshake* (see Further Reading, page 111).

of non-locality that we encountered earlier, but, of course, that was not in Einstein's mind in 1909.

One of the few people to take the idea seriously was Richard Feynman, when he was a research student at Princeton in the 1940s. Encouraged by his thesis advisor, John Wheeler,* he developed the idea that when an electron interacts with another charged particle a kind of half wave goes out into both the future and the past. Where the wave meets another charged particle, the other particle produces its own half wave going both forwards and backwards in time. But in Feynman's version of the theory, the two half waves interfere to cancel each other out everywhere except in the space between the two particles, where they reinforce to make a full wave. When he gave a talk on the topic at Princeton, among the luminaries in the audience were Einstein and Wolfgang Pauli. Pauli said that he did not think the idea would work, and asked Einstein if he agreed. 'No', said Einstein, 'I find only that it would be very difficult to make a corresponding theory for gravitational interaction.'

In spite of this endorsement, the idea languished, because people simply did not 'believe in' waves coming from the future. But in the late 1970s, while teaching at the University of Washington in Seattle, John Cramer, who had been intrigued by Feynman's idea since coming across it as a graduate student twenty years earlier, had a flash of insight into how the idea could be incorporated in quantum mechanics. Like many good ideas, it is obvious once somebody has pointed it out.

....

* The same John Wheeler. He had a long and impressive track record.

Cramer's insight was triggered by thinking about what happens to the 'probability wave' in a quantum system when the particle that it is associated with is detected at a definite location. How does the wave everywhere else 'know' to vanish at that instant? He makes an analogy with a bottle thrown into the Atlantic Ocean from a beach in Florida. Imagine that this is a quantum bottle that disappears into a wave which spreads out across the ocean to Europe. On a beach in England, the bottle appears. At that instant, the waves spread across the entire ocean disappear. Cramer realised that there must be advanced and retarded waves having quantum 'handshakes' all over the place, and that only those retarded waves that made advanced-wave 'echoes' could affect the location of particles – their mysterious quantum-mechanical transfer from A to B (or from one energy level to another) without passing through the space in between. Waves from the bottle in England had travelled backwards in time to Florida, and across the ocean, to establish a unique connection and cancel the other waves out. To Cramer, this looked a lot like the pilot wave model, which has waves showing the particles where to go, but, crucially, does not have the time-reversed confirmation of the handshake.

This also explains the EPR puzzle. Two particles that have once interacted are each connected, ever afterwards, by handshakes between them and the site of their interaction. All of this ties in with the correct (in Cramer's view) description of Schrödinger's famous equation.*

....

* The following section is adapted from my book *Schrödinger's Kittens*.

In order to apply the absorber theory ideas to quantum mechanics, you need a quantum equation, which, like Maxwell's equations, yields two solutions, one equivalent to a positive energy wave flowing into the future, and the other describing a negative energy wave flowing into the past. At first sight, Schrödinger's equation doesn't fit the bill, because it only describes a flow in one direction, which (of course) we interpret as from past to future. But as all physicists learn at university (and most promptly forget) the most widely used version of this equation is incomplete. As the quantum pioneers themselves realised, it does not take account of the requirements of relativity theory. In most cases, this doesn't matter, which is why physics students, and even most practising quantum mechanics, happily use the simple version of the equation. But the full version of the wave equation, making proper allowance for relativistic effects, is much more like Maxwell's equations. In particular, it has two sets of solutions – one corresponding to the familiar simple Schrödinger equation, and the other to a kind of mirror image Schrödinger equation describing the flow of negative energy into the past.

This duality shows up most clearly in the calculation of probabilities in the context of quantum mechanics. The properties of a quantum system are described by a mathematical expression, called the state vector, which is described by Schrödinger's wave equation. In general, this is a complex number. A complex number is one which involves the square root of minus 1, which is written as i. So if a and b are everyday numbers, $(a + ib)$ would be a complex number, and so would $(a - ib)$. The probability calculations needed to work out the chance of finding an electron (say) in a particular place at a particular time actually depend on calculating the square of the state vector corresponding to that particular state of the electron.

But calculating the square of a complex variable does not simply mean multiplying it by itself. Instead, you have to make another variable, a mirror image version called the complex conjugate, by changing the sign in front of the imaginary part: if it was + it becomes –, and vice versa. So $(a - ib)$ is the complex conjugate of $(a + ib)$. The two complex numbers are then multiplied together to give the probability. But for equations that describe how a system changes as time passes, this process of changing the sign of the imaginary part and finding the complex conjugate is equivalent to reversing the direction of time! The basic probability equation, developed by Max Born back in 1926, itself contains an explicit reference to the nature of time, and to the possibility of two kinds of Schrödinger equations, one describing advanced waves and the other representing retarded waves.

The remarkable implication is that ever since 1926, every time a physicist has taken the complex conjugate of the simple Schrödinger equation and used it to calculate a quantum probability, they have actually been taking account of the advanced wave solution to the equations, and the influence of waves that travel backwards in time, without knowing it. There is no problem at all with the mathematics of Cramer's interpretation of quantum mechanics, because the mathematics, right down to Schrödinger's equation, is exactly the same as in the Copenhagen Interpretation. The difference is, literally, only in the interpretation.

The way Cramer describes a typical quantum transaction is in terms of a particle shaking hands with another particle somewhere else in space and time. He started from the idea of an electron emitting electromagnetic radiation which is absorbed by another electron, but the description works just as well for the state vector of a quantum entity which starts out in one state and ends up in another state as a result of an interaction – for example, the state vector of a particle emitted from a source on one side of the experiment with two holes and absorbed by a detector on the other side of the experiment.

One of the difficulties with any such description in ordinary language is how to treat interactions that are going both ways in time simultaneously, and are therefore occurring instantaneously as far as clocks in the everyday world are concerned. Cramer does this by effectively standing outside of time, and using the semantic device of a description in terms of some kind of pseudotime. This is no more than a semantic device – but it certainly helps most people to get the picture straight in their mind.

It works like this. When a quantum entity (the emitter) interacts with the outside world, on this picture, it attempts to do so by producing a field which is a time-symmetric mixture of a retarded wave propagating into the future and an advanced wave propagating into the past. As a first step in getting a picture of what happens, ignore the advanced wave and follow the story of the retarded wave. This heads off into the future until it encounters an entity (the absorber) with which it can interact. The process of interaction involves making the second entity produce a new retarded field which exactly cancels out the first retarded field. So in the future of the absorber, the net effect is that there is no retarded field.

But the absorber also produces a negative advanced wave travelling backwards in time to the emitter, down the track of the original retarded wave. At the emitter, this advanced wave is absorbed, making the original entity recoil in such a way that it radiates a second advanced wave back into the past. This 'new' advanced wave exactly cancels out the 'original' advanced wave, so that there is no effective radiation going back in the past before the moment when the original emission occurred. All that is left is a double wave linking the emitter and the absorber, made up half of a retarded wave carrying positive energy into the future and half of an advanced wave carrying negative energy into the past (in the direction of negative time).

Because two negatives make a positive, this advanced wave adds to the original retarded wave as if it too were a retarded wave travelling from the emitter to the absorber. Negative energy and negative time add up to make positive energy going forward in time. In Cramer's words:

> The emitter can be considered to produce an 'offer' wave which travels to the absorber. The absorber then returns a 'confirmation' wave to the emitter, and the transaction is completed with a 'handshake' across spacetime.*

But this is only the sequence of events from the point of view of pseudotime. In reality, the process is atemporal; it happens all at once.

'If there is one particular link in [the] event chain that is special', says Cramer, 'it is not the one that ends the chain. It is the link at

---

* *Reviews of Modern Physics*, Volume 58, p. 647, 1986.

the beginning of the chain when the emitter, having received various confirmation waves from its offer wave, reinforces one of them, chosen at random in accordance with the rules of probability, in such a way that it brings that particular confirmation wave into reality as a completed transaction. The atemporal transaction does not have a "when" at the end.'

How does this resolve the central mystery of the experiment with two holes? According to the TI, a retarded 'offer wave' spreads out through both holes in the experiment, and triggers an advanced 'confirmation wave' from the detector screen which travels back through both holes in the experiment to the source. Each particle chooses which offer to accept at random, producing an interference pattern. But if, in a cunning delayed choice version of the experiment, one of the holes is blocked off after the particle has set out on its journey, the particle already 'knows' this, because the confirmation wave only had one hole to go through as it went back to make the handshake. Cramer:

> The issue of when the observer decides which experiment to perform is no longer significant. The observer determined the experimental configuration and boundary conditions, and the transaction formed accordingly. Furthermore, the fact that the detection event involves a measurement (as opposed to any other interaction) is no longer significant, and so the observer has no special role in the process.

This success in resolving the puzzles of quantum physics has been achieved at the cost of accepting just one idea that seems to run counter to common sense – the idea that part of the quantum wave really can travel backwards through time. At first sight, this is in stark disagreement with our intuition that causes must always precede the events that they cause. But on closer inspection it turns out that the kind of time travel required by the Transactional Interpretation does not violate the everyday notion of causality after all. When an atemporal handshake takes place with the aid of an advanced quantum wave that travels backwards in time, this does not have any influence on the logical pattern of causality in the everyday world.

It should be no surprise that the way the Transactional Interpretation deals with time differs from common sense, because the Transactional Interpretation explicitly includes the effects of relativity theory. The Copenhagen Interpretation, by contrast, treats time in the classical, 'Newtonian' way, and this is at the heart of the inconsistencies in any attempt to explain the results of quantum experiments measuring Bell's Inequality in terms of the Copenhagen Interpretation. If the velocity of light were infinite, the problems would disappear; there would be no difference between the local and non-local descriptions of processes involving Bell's Inequality, and the ordinary Schrödinger equation would be an accurate description of what is going on – the ordinary Schrödinger equation is, in effect, the correct 'relativistic' equation when the speed of light is infinite.

How does the atemporal handshaking affect the possibility of free will? At first sight, it might seem as if everything is fixed by these communications between the past and the future. Every photon that is emitted already 'knows' when and where it is going to be absorbed; every quantum probability wave, slipping at the speed of light through the slits in the experiment with two holes, already 'knows' what kind of detector is waiting for it on the other side. We are faced with the image of a frozen Universe, in which neither time nor space have any meaning, and everything that ever was or ever will be just is.

But in my time-frame decisions are made with genuine free will and no certain knowledge of their outcomes. It takes time (in the macroscopic world) to make the decisions (both human decisions and quantum 'choices' like those involved in the decay of an atom) which make the atemporal reality of the microscopic world.

Cramer is at pains to stress that his interpretation makes no predictions that are different from those of conventional quantum mechanics, and that it is offered as a conceptual model which may help people to think clearly about what is going on in the quantum world, a tool which is likely to be particularly useful in teaching, and which has considerable value in developing intuitions and insights into otherwise mysterious quantum phenomena. But there is no need to feel that the Transactional Interpretation suffers in comparison with other interpretations in this regard, because none of them is anything other than a conceptual model designed to help our understanding of quantum phenomena, and all of them make the same predictions.

Therein lies the rub. All the Solaces are equally good; all of them are equally bad. At least that means you are free to choose whichever one gives you most comfort, and ignore the rest.

CONCLUSION

There Ain't No Sanity Clause

For the past ninety years, many of the best scientific brains on Earth have puzzled over the meaning of quantum mechanics. The six possible Solaces I have described here are the best ideas they have come up with, and they can be summed up briefly:

One. The world does not exist unless you look at it.

Two. Particles are pushed around by an invisible wave, but the particles have no influence on the wave.

Three. Everything that could possibly happen does, in an array of parallel realities.

Four. Everything that could possibly happen already has happened and we only noticed part of it.

Five. Everything influences everything else instantly, as if space did not exist.

Six. The future influences the past.

As Feynman wrote in *The Character of Physical Law*: 'I think I can safely say that nobody understands quantum mechanics ... Do not keep saying to yourself, if you can possibly avoid it, "But how can it be like that?" because you will go "down the drain" into a blind alley from which nobody has yet escaped. Nobody knows how it can be like that.'

FURTHER READING

Easier stuff

Philip Ball, *Beyond Weird* (Bodley Head, London, 2018)

Brian Clegg, *The Quantum Age* (Icon Books, London, 2014)

John Gribbin, *The Quantum Mystery* (Kindle Single)

David Lindley, *Where Does The Weirdness Go?* (Basic Books, New York, 1996)

George Musser, *Spooky Action at a Distance* (*Scientific American/ Farrar, Strauss & Giroux*, New York, 2015)

Heinz Pagels, *The Cosmic Code* (Michael Joseph, London, 1982)

Euan Squires, *The Mystery of the Quantum World*, second edition (Institute of Physics, Bristol, 1994)

Middling stuff

John Cramer, *The Quantum Handshake* (Springer, Heidelberg, 2016)

Richard Feynman, *The Character of Physical Law*, new edition (Penguin, London, 1992)

Harder stuff

John Bell, *Speakable and Unspeakable in Quantum Mechanics* (Cambridge University Press, 1987)

Richard Feynman, Robert Leighton and Matthew Sands, *The Feynman Lectures on Physics*, Volume III (Addison-Wesley, Reading, MA, 1965)

Leonard Susskind and Art Friedman, *Quantum Mechanics* (Allen Lane, London, 2014) (very hard, but very thorough)

Funny stuff

https://www.poetryfoundation.org/poems/43909/the-hunting-of-the-snark

SEVEN

PILLARS OF

SCIENCE

The Incredible Lightness
of Ice, and Other
Scientific Surprises

CONTENTS

LIST OF ILLUSTRATIONS

Wisdom hath builded her house, she hath hewn out her seven pillars.

Proverbs 9:1

PREFACE

Seven Pillars of Wisdom

J.B.S. Haldane famously described the four stages of acceptance for scientific ideas as:

i)   this is worthless nonsense;

ii)   this is an interesting, but perverse, point of view;

iii) this is true, but quite unimportant;

iv)   I always said so.

The more I look at the history of science – and the longer I observe the ongoing development of science – the more I appreciate the truth of this aphorism. Looking back, it is easy to see how ideas that were once outrageous became accepted truths, and to feel a sense of superiority over those simpletons who, for example, thought that the Earth was flat. But even in my own lifetime I have seen ideas once regarded as wild speculations – including the Big Bang theory of the origin of the Universe and the non-locality of quantum entities – become received wisdom, pillars of science, while more 'commonsensical' alternatives – the Steady State theory, the idea that what happens in one location cannot instantly affect what happens somewhere far away – have fallen by the wayside. How science works is as fascinating as the science itself, and to demonstrate this I have picked out seven examples which were each sensational in their day, and which have either become pillars of scientific wisdom or are well on their way to passing through Haldane's four stages of acceptance. In order to restrict myself to seven, I needed some overall theme to link them, and I have chosen features of the Universe which are closely related to our own existence, and to the possibility of life elsewhere. This is, after all, the most important aspect of science as far as we humans are concerned.

Some of these examples are already pillars of science, others may be at an earlier stage – I leave you to judge which ones. But although all were sensational in their day, and some still are, a key feature of the development of science is a willingness to think the unthinkable, and then, crucially, to test those ideas and find out if they are good descriptions of what is going on in the real world. There are, though, some ideas which are impossible to categorise, and which, depending on your personal point of view, might be assigned to any one of Haldane's stages. The biggest of these is a question that has puzzled philosophers for much longer than what we call science has existed, and with which I shall top and tail this book – are we alone in the Universe?

PROLOGUE

Worlds Beyond:
Maybe We Are Not Alone

The Earth is round and moves through space. This was a dramatic realisation only a few hundred years ago. It flies in the face of common sense, so much so that some people still cannot accept it. You may not be one of those people, but do you just accept the story because it is what you were told as a child and 'everybody knows' it is true? Or have you ever stopped to think what a crazy idea this is, in terms of your everyday experience, and to consider the evidence?

To see how reasonable the idea of a flat Earth is, and how sensational was the realisation that it is round, we can look back to the Greek philosopher Anaxagoras of Athens, who was around in the fifth century BCE. Anaxagoras was no fool. He based his reasoning on the best evidence available to him, and given those facts his reasoning was correct. His conclusions turn out to have been wrong, but far more important than that is the fact that he tried to understand the Sun as a physical entity subject to the same laws as those that apply to things here on Earth. He did not treat it as a supernatural phenomenon beyond human comprehension.

The trigger for Anaxagoras' speculations was a meteorite which fell one day at Aegospotami. The meteorite was hot, so he inferred that it must have come from the Sun. It contained iron, so he inferred that the Sun must be made of iron – a hot ball of iron travelling across the sky. All this was completely logical in the light of the state of knowledge at the time. But it raised two intriguing questions that Anaxagoras set out to answer – how big must that ball of hot iron be, and how far above the surface of the Earth was it moving?

Anaxagoras wasn't much of a traveller, but he had heard accounts from people who had been to the Nile delta, and beyond to the upper reaches of the Nile. These accounts mentioned that at the stroke of noon on the summer solstice (the 'longest day'), the Sun was vertically overhead at a city called Syene, near the present-day location of the Aswan dam. You may have come across this tit-bit of information in another context; if so, be prepared for a surprise. Anaxagoras also knew that on the longest day at noon the Sun was at an angle of 7 degrees out of the vertical at the Nile delta. And he knew the distance from the delta to Syene. With this information, assuming the Earth was flat and using the geometry of right-angled triangles, it was a trivial matter for Anaxagoras to calculate that at noon on the summer solstice the Sun was roughly 4,000 miles (in modern units) above the heads of the inhabitants of Syene. Because the Sun covers roughly half a degree of arc on the sky (the same as the Moon, a dramatic coincidence outside the scope of this book), the geometry of triangles also told him that it must be about 35 miles across, roughly the same as the southern peninsula of Greece, the Peloponnesus.

The suggestion that the Sun was a natural phenomenon was so shocking to his fellow citizens that Anaxagoras was arrested for heresy, and banished for ever from his home city of Athens. It would be more than two thousand years, not until the seventeenth century, before another thinker, Galileo Galilei, also tried to explain the Sun as a natural phenomenon, and was also accused of heresy.

But only a couple of hundred years after Anaxagoras another Greek philosopher, Eratosthenes, used exactly the same data in a slightly different calculation. This is the version of the story you may have heard. Eratosthenes assumed that the Earth is spherical, and guessed that the Sun is so far away that rays of light from the Sun reach the Earth along parallel lines. With this assumption, the angle of 7 degrees out of the vertical measured at the Nile delta is the same as the angle subtended at the surface of the Earth by the distance from the delta to Syene, measured from the centre of the Earth (see diagram overleaf). This makes it possible to calculate the radius of the Earth. Because the angle is the same, the 'answer' is the same – 4,000 miles. But this is now interpreted as the radius of the Earth, not the distance of the Sun above the Earth. Because Eratosthenes was 'right', his is the version of the story recorded in textbooks and popular accounts, while Anaxagoras is ignored. But the moral is not who was right and who was wrong. Good theories are based on sound evidence and make predictions that can be tested. If the theory passes those tests, it continues to be used; if it fails those tests it is rejected. Taken together, the two theories (strictly speaking, hypotheses, but I won't quibble) of the Greek philosophers combine to tell us that either the Earth is flat and the Sun is about 4,000 miles above it, or the Earth is a ball with a radius

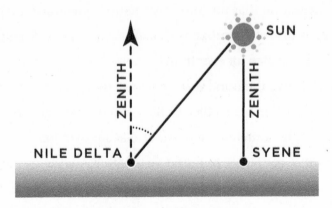

Assuming the Earth is flat,
it is simple to calculate the distance to the Sun

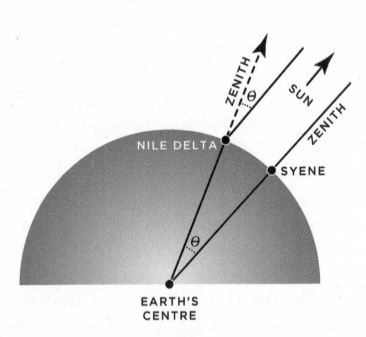

Assuming the Earth is round,
the same observation tells us the radius of the planet

of about 4,000 miles and the Sun is at a vast but unknown distance. Later observations and measurements made it possible to decide which is a better description of the real world.

There is also a cautionary aspect to the tale. Even a radical and far-sighted thinker who was not afraid to confront the authorities of his day in his quest for the truth could not rid himself of the preconception that the Earth is flat. Anaxagoras never considered alternatives. The history of science is filled with similar unfortunate examples of ideas that are built up with impeccable logic and complete accuracy, but are based on unquestioning faith in something which turns out to be completely untrue. Science should not be about faith, but about questioning cherished beliefs. Not that this always makes for a quiet life, as Giordano Bruno found to his cost. Mind you, Bruno seems to have gone out of his way to make his life difficult, and not just in the pursuit of science.

Historians (including myself) often date the beginning of modern science to the publication in 1543 of the book *De Revolutionibus Orbium Coelestium* (*On the Revolution of the Celestial Spheres*), by Nicolaus Copernicus. In truth, though, the book was not a sensation at the time; the ideas it contained did not gain widespread currency for the best part of a hundred years, and it did not go far enough in displacing ourselves from the centre of the Universe. Copernicus retained the idea that there is a fixed centre to the Universe, but moved this from the Earth to the Sun. He explained the apparent movement of the stars across the sky as due to the rotation of the Earth, but retained the idea that the stars and planets were fixed to solid spheres moving around the Sun. His most 'heretical'

Giordano Bruno
Science Photo Library

suggestion was that the Earth also is a planet, orbiting the Sun once a year, but that is as far as he went.

Bruno picked up Copernicus' ball and ran off with it. Born near Naples in 1548, five years after the publication of *De Revolutionibus*, and christened Filippo, Bruno joined the Dominican order at the age of seventeen, taking the name Giordano, and became an ordained priest in 1572. He soon ran into difficulties because of his free thinking and taste for forbidden (or at least, controversial) books. He seems to have got into particular trouble for espousing Arianism, the belief that Jesus occupies a position intermediate between Man and God, making him divine but not the same as God. When things became too hot, he fled Naples, discarded his religious garments, and began a series of wanderings that took him to, among other places, Geneva, Lyon and Toulouse, where he took a doctorate in theology and lectured on philosophy. In 1581 he moved to Paris, where he enjoyed the safety of the protection of the king, Henry III, and published several works.

In 1583 Bruno went to England with letters of recommendation from the French king, and moved in Elizabethan court circles where he met such notable people as Philip Sidney and (possibly) John Dee. Although he gave some lectures in Oxford on the Copernican model of the Universe, he was unable to obtain a position at the university, where his controversial views were derided by John Underhill, then Rector of Lincoln College and later Archbishop of Canterbury, who sneered at Bruno for espousing 'the opinion of Copernicus that the earth did go round, and the heavens did stand still; whereas in truth it was his own head which rather did run round, and his brains did not stand still'.[1] It appears, though, that it was as much Bruno's personality as his teaching that made him unwelcome in Oxford. He seems to have been arrogant and unwilling to give much time to people he regarded as fools, and managed to put up the backs even of people who shared his views.

But this was less than half of what Bruno proposed. In 1584 he published two of a series of 'dialogues' in which he supported the Copernican cosmology, and by 1588 he was writing that the Universe is 'infinite … endless and limitless'. So what were the stars? Pulling together the ideas expressed by Bruno in several places, he was the first person to realise that not only are the stars other suns, but that like the Sun itself they could each have their own family of planets. These other worlds, he said, 'have no less virtue nor a nature different from that of our Earth', and therefore they could 'contain animals and inhabitants'.

This would have been enough to bring him into further conflict with the Roman Catholic authorities, and Bruno is sometimes held up as a martyr for science. But his problems with the authorities ran so deep that these beliefs actually amount to no more than a footnote in the story of his later life and fate. In 1585, because of a deteriorating political situation between England and France, Bruno returned to Paris, then on to Germany and Prague, where he achieved the distinction (having already fallen foul of the Catholic authorities) of being excommunicated by the Lutherans. In 1591 he took a chance on returning to Italy, initially to Padua in the hope of getting a professorship. But the job went to Galileo and he moved on to Venice, the most liberal of the Italian city states. Not liberal enough, as it turned out. On 22 May 1592 Bruno was arrested and charged with blasphemy and heresy, his belief in the plurality of worlds just one of many examples in the citation. He might have got away with a relatively light sentence, but the Inquisition demanded that he should be transferred to Rome for them to deal with, and the Venetian authorities eventually bowed to pressure and handed him over in February 1593.

Bruno's trial lasted for seven years, off and on, during which time he was imprisoned in Rome. Many of the papers relating to the trial have been lost, but the charges against him included not only broad-brush blasphemy and heresy but immoral conduct. Specific charges are thought to have included speaking and writing against the idea of the Trinity and the divinity of Christ, and doubting the virginity of Mary, mother of Jesus. He also made the shocking suggestion that different branches of the Christian Church should live in harmony and respect each other's views. These were much greater sins in the eyes of the Inquisition than speculating about the plurality of worlds, but that went on the list anyway.* As usual with heretics, Bruno was eventually given an opportunity to recant, which he refused, and on 20 January 1600 Pope Clement VIII formally declared him a heretic. He is alleged to have made a threatening gesture at the judges when sentenced; he was burned at the stake on 17 February 1600, having first been gagged to prevent any heretical last words being heard by the onlookers. So here are some of his not-quite-last words which demonstrate the breadth of his thinking:

....

* Arguably, if it hadn't been for Bruno the Church might not have got so worked up about Copernicus – his book was only placed on the Index of banned works in 1616, and remained there until 1835.

There is no absolute up or down, as Aristotle taught; no absolute position in space; but the position of a body is relative to that of other bodies. Everywhere there is incessant relative change in position throughout the universe, and the observer is always at the centre of things.

Although it quite quickly became appreciated that the stars are indeed other suns – Isaac Newton was one of several people who tried to estimate the distances to stars by assuming that they had roughly the same brightness as our Sun – it was only in the 1840s that astronomers were able to measure a few of those distances directly using the geometrical technique of parallax, which uses the shift in position of nearby stars against the background of distant stars as the Earth moves around its orbit. It was only in the twentieth century that other techniques made it possible to measure distances far out across the Universe, and eventually, by the 1930s, to make the notion of an infinite Universe respectable. But even then the idea that the stars might have their own families of planets remained pure speculation.

The situation changed in 1995, with the discovery of a planet orbiting a star, labelled 51 Pegasi, roughly similar to the Sun. The discovery was made by analysing the wobble of the star caused by the gravitational pull of the planet orbiting around it. These measurements are possible because the wobble causes a tiny shift in the spectral lines of the star,* a process known as the Doppler effect. The measurements turned out to be relatively easy, because the planet is very big and

....

* Stellar spectroscopy is one of the pillars discussed later.

orbits rather close to the star, so it has a relatively large gravitational influence. This was not what astronomers were expecting.

In our Solar System, there are four small rocky planets (very roughly like the Earth) orbiting in the inner region nearer the Sun, and four large gaseous planets (very roughly like Jupiter) orbiting in the outer regions, plus various small bits of debris including the object Pluto. With no other information to go on, astronomers guessed that other planetary systems might be similar. But the planet found orbiting 51 Pegasi is very large and orbiting very close to its star. It became known as a 'hot jupiter'. It has more than half the mass of Jupiter, the largest planet in our Solar System, and orbits its star at only a tenth of the distance of Mercury, the innermost planet in the Solar System, from our Sun. The first lesson drawn from this is that you cannot generalise on the basis of one example! Our Solar System is clearly not the only kind of planetary system in the Universe, and may even be unusual. And the corollary is that we should not assume that the Earth is a typical planet; more on this later.

Since 1995 many more 'extrasolar' planetary systems have been discovered, many of them harbouring hot jupiters, and many now known to have multiple planets, in a variety of configurations, orbiting the central stars. The discovery of a 'new' planet is no longer news, let alone headline news, unless it is what news outlets like to call an 'Earth-like' planet. But be wary of the headlines. All that they mean by this is that the planet is probably rocky and has a few times the mass of the Earth. It is Earth-sized, rather than Earth-like. And to clarify the distinction we have only to look at our nearest neighbour in the Solar System, the planet Venus, orbiting a little closer to the Sun than us. Venus is almost exactly the same size as the Earth, it is rocky, and overall it is a better candidate for the description 'Earth-like' than any of the extrasolar planets trumpeted in the media. It has a similar size, mass, density and surface gravity. But the temperature on the surface of Venus is 462°C, hot enough to melt lead. This temperature is not because the planet is slightly closer to the Sun than we are, but thanks to a strong greenhouse effect produced by its thick carbon dioxide atmosphere. The atmospheric pressure on the surface of Venus is 92 times the pressure at the surface of Earth, the same as the pressure a kilometre below the surface of our oceans.

Which brings us back to Bruno and his suggestion that there is a multitude of planets populated by a multitude of life forms, including people. Planets there are, in profusion. Every Sun-like star, and perhaps every star, seems to have a family of planets. Let's not be too pessimistic. There are hundreds of billions of stars in the Milky Way, our home galaxy, a kind of island in the Universe. Even if only a small proportion of them have planetary systems like our Solar System, and even if only a small proportion of those systems have at least one planet which is more Earth-like than Venus-like, there could be millions of potential homes for life forms like us out there, before we even consider more exotic possibilities. One per cent of 100 billion is still a billion, and one per cent of a billion is 10 million. Planets are common in the Universe – maybe even planets like Earth. Maybe we are not alone. Homes for life may be common. But what about life itself? How did we get here? The answer depends on seven surprising discoveries about the way the Universe works – seven pillars of science that underpin our existence, and the possible existence of other life in the Universe.

1

Solid Things Are Mostly Empty Space

S olid objects are empty. Although this is an often cited example of the non-commonsensical nature of the world, it still brings you up short if you stop to think about it. Things like the 'solid' keyboard I am typing on and the fingers doing the typing are made up of tiny particles spread through relatively huge volumes of space, held together by electric forces. This is such an important and mind-blowing idea that Richard Feynman said it was the most significant fact that science had discovered about the world. As on so many topics, it is worth quoting him verbatim:

> If, in some cataclysm, all of scientific knowledge were to be destroyed, and only one sentence passed on to the next generation of creatures, what statement would contain the most information in the fewest words? I believe it is the *atomic hypothesis* (or the atomic *fact*, or whatever you wish to call it) that *all things are made of atoms – little particles that move around in perpetual motion, attracting each other when they are a little distance*

apart, but repelling upon being squeezed into one another. In that one sentence, you will see, there is an enormous amount of information about the world, if just a little imagination and thinking are applied.[2]

Few physicists, however, have the power of imagination (or better, physical insight) and thinking that Feynman had, and the debate about whether the world is really made of such particles was not resolved until the early years of the twentieth century, although the idea of atoms had been suggested much earlier.

Popular accounts of the atomic theory (or whatever you wish to call it) usually start out with a nod to Democritus, who lived in the fifth century BCE, and Epicurus, who was around between about 342 BCE and 271 BCE. But their idea of little objects moving about in 'the void' and interacting with one another was never more than a minority opinion, ridiculed by philosophers such as Aristotle who rejected the idea of a void. It wasn't until 1649 that Pierre Gassendi revived the idea and suggested that atoms had different shapes and could join together through a kind of hook-and-eye mechanism. He stressed that there was nothing at all in the gaps between atoms. This was the beginning of a debate that rumbled on for more than two hundred years. On one side there was what we might call the Newtonian school of thought, after Isaac Newton, which favoured the atomic hypothesis; on the other, the Cartesian school, after René Descartes, who abhorred the idea of a void, or vacuum. Things came to a head in the nineteenth century.

From the 1850s onward, building on the earlier work of John Dalton, chemists increasingly accepted the idea of atoms, with atoms of different elements having different weights, and joining together to make molecules, so that a molecule of water, for example, was regarded as a combination of two hydrogen atoms with one oxygen atom. They could measure the weights (strictly speaking, masses) of atoms of different elements compared with that of hydrogen, the lightest element. And they were even able to calculate how many particles (atoms or molecules) there must be in a sample of any element that contained its atomic (or molecular) weight in grams – 1 gram of hydrogen, 12 grams of carbon, 16 grams of oxygen, and so on. Each such sample would have the same number of particles. This number became known as Avogadro's number, after the pioneer who developed the theory behind it, and it is very big. But before I go into how big it is, I should spell out the opposition to these ideas that persisted even at the beginning of the twentieth century, and which highlights how sensational the idea of atoms really is.

The opposition came from physicists and philosophers who pointed out what they saw as a fatal flaw in the idea of large numbers of tiny particles moving around in empty space, bouncing off each other and going merrily on their way in accordance with the laws of motion spelled out by Isaac Newton. The relevant thing about Newton's laws is that they are reversible. The standard way of highlighting this is to think of a collision between two pool balls. One ball moves in from, say, the left, hits a stationary ball and stops, while the other ball moves off to the right. If you made a movie of this event and ran it backwards, it would still look entirely OK. A ball would move in from the right, collide with a stationary ball and stop while the other ball moved off to the left. Newton's laws do not contain an 'arrow of time'. But the real world does have a direction of time built in to it. If we now imagine the cue ball striking the pack of pool balls in a break so that they scatter in all directions, the situation is not reversible, even though every single collision between the balls obeys Newton's laws. 'Running the movie backwards' produces a sequence never seen in the everyday world – balls arriving from all directions, colliding and settling into a neat pack while just one ball zooms off towards the cue.

The irreversibility of the everyday world was expressed by nineteenth-century scientists in terms of heat – the science of thermodynamics. They pointed out that heat always flows from a hotter object to a colder one. An ice cube placed in a glass of warm water gains heat from the water and melts; we never see water in a glass spontaneously getting warmer while a lump of ice forms in the middle. But both this scenario and the 'reversed' pool ball break are entirely allowed by Newton's laws. The initial conclusion of the nineteenth-century thermodynamicists was that things could not really be made of tiny particles bouncing around in accordance with those laws. But then the dilemma was resolved.

No fewer than three great thinkers independently found the solution. They realised that the behaviour of large numbers of particles interacting in accordance with Newton's laws had to be described in statistical terms, and they worked out the equations to calculate how very large numbers of particles would behave – the laws of what became known as statistical mechanics. This tells us, in a rigorous mathematical way, that although there is nothing in the laws of physics to prevent ice cubes forming in glasses of warm water, such an event is extremely unlikely, and will only occur once in a very, very long time – a time which can be calculated if you know how many particles are involved.* The first two scientists to appreciate this and work out the laws of statistical mechanics can be excused for not knowing about each other's work. Ludwig Boltzmann worked in Europe, while Willard Gibbs worked in the USA, and even at the turn of the nineteenth century scientific ideas took a while to cross the Atlantic. The third inventor (or discoverer) of statistical mechanics had less excuse, not least since he came on the scene a little later. But he was notorious for not bothering to keep up with what other people were doing, preferring to work everything out for himself. His name was Albert Einstein, and it is a sign of how the atomic theory of matter had failed to become established that at the beginning of the twentieth century he set out to find evidence 'which would guarantee as much as possible the existence of atoms of definite finite size'.[3] His version of statistical mechanics appeared in a series of three extraordinary papers, published between 1902 and 1904, which would have assured him of scientific fame, if only he had been first on the scene. But in 1905, among other things he did produce the scientific paper which finally established the reality of atoms and molecules to all but a few die-hard philosophers. It's also much easier for non-mathematicians to grasp, so I shall cast statistical mechanics to one side and focus on the physics.

The physics harks back to an old piece of work which Einstein was at least aware of, but only in a vague sort of way. And it wasn't the jumping-off point for his own work, because once again he was

---

* You would have to sit watching that glass of water for vastly longer than the present age of the Universe to have much chance of seeing an ice cube form.

working it out from first principles, this time trying to calculate how a small piece of material – such as a dust grain – suspended in a liquid – such as a glass of water – would move as it was buffeted about by atoms and molecules striking it from all sides. This kind of motion had been studied by the Scottish botanist Robert Brown back in the 1820s. His interest stemmed from observations, made using microscopes, of pollen grains dancing about in water in a jittery kind of motion, like running on the spot. The natural explanation at the time was that the pollen grains were alive, and moving under their own steam. But Brown tested this by looking at grains of ground-up glass and granite in water, and found that they danced in the same way. This established that the dancing had nothing to do with life, and it became known as Brownian motion.

Einstein started out by calculating how atoms and molecules would make inanimate dust grains move in a liquid, but starting from the bottom up, not from the top down. In the first paragraph of the paper on the subject he produced in 1905, he says:

> It is possible that the motions to be discussed here are identical with so-called Brownian molecular motion; however, the data available to me on the latter are so imprecise that I could not form a judgement on the question.

The 'data available' were 'so imprecise' because he couldn't be bothered to look them up; and there must be a strong suspicion that this sentence was added after some friend read a draft of the paper and pointed out to him the link to Brownian motion. But whatever his motivation, Einstein explained Brownian motion with one of those pieces of insight that geniuses come up with, but which then make you wonder why nobody else thought of it, backed up by calculations which gave the experimenters something to test.

Particles large enough to be seen using contemporary microscopes – grains like pollen, or ground-up glass – were, Einstein realised, far too small to be moved visibly by the impact of a single atom or molecule. But in a liquid, such particles are constantly being bombarded on all sides by large numbers of atoms and molecules. This bombardment cannot be perfectly even. At any instant, a few more impacts will occur on one side, and a few less on another. The particle will shift a little in the direction of fewer impacts. But then the balance will change, and it will be nudged in a different direction. The overall effect is that it jitters about, not quite running on the spot but jogging in a zigzag path and gradually getting further away from where it started. The path is now known as a random walk; and this was Einstein's key insight.

Einstein showed that wherever the particle starts from, the distance it moves away from that point depends on the square root of the time that has passed. So if it moves a certain distance in one second it will move twice as far in four seconds (because 2 is the square root of 4), four times as far in sixteen seconds, and so on. But it doesn't keep going in the same direction. After four seconds it is twice as far away, but in a random and unpredictable direction; after sixteen seconds it is four times as far away in another random direction. This is called 'root mean square' displacement, and it was possible for experimenters to test the prediction. Plugging in Avogadro's number from other studies, Einstein concluded that a particle with a diameter of 0.001mm in water at 17°C would shift position by six millionths of a metre from its starting point in one minute. The modern calculation of Avogadro's number, the number of molecules in the molecular weight of a substance in grams, is equal to $6.022140857 \times 10^{23}$, or roughly a 6 followed by 23 zeroes. This gives you some idea of why the statistical behaviour of matter overwhelms the individual reversible interactions to produce effects like melting ice cubes and Brownian motion.* As Einstein summarised:

> If the prediction of this motion were to be proved wrong, this fact would be a far-reaching argument against the molecular-kinetic conception of heat.

Of course it was not proved wrong, and this was taken as clinching evidence of the reality of atoms and molecules. But there's more – more even than Einstein realised in 1905.

The molecular-kinetic theory of heat that Einstein mentioned explains the division of everyday things into solid, liquid, or gaseous states. A gas is the archetypal example of atoms moving in the void, with nothing between them. A liquid is envisaged as a collection of atoms (or molecules) sliding past one another fairly freely, with no space between them. And in a solid, the particles are pictured as set firmly in an array, touching one another, again with no spaces

----

* In case you are worried about those pool balls, in order for a set of stationary balls lying on the table to start moving together into a pack, the material of the table would have to cool down as it gave up energy to the balls, like ice giving up energy to the water in a glass. This is possible, but extremely unlikely, because of the vast number of particles in the table that would have to work together, *not* because of the relatively small number of pool balls having to work together.

between the atoms or molecules. So why did I describe my keyboard and my fingers as mostly empty space? This was a really sensational discovery, and it was made by researchers in Manchester at the end of the first decade of the twentieth century, little more than a hundred years ago.

The people who actually did the experiments were Hans Geiger and Ernest Marsden, working under the supervision of Ernest Rutherford. Rutherford was one of the key figures in the development of physics around this time. He came from New Zealand, and in the 1890s worked in Cambridge, England, where he investigated the behaviour of the newly-discovered X-rays, then in 1898 moved on to McGill University in Montreal where he investigated the other great discovery of the time, radioactivity. He settled in Manchester in 1907. Within a year, his team had established that one form of this radiation, called alpha radiation, is actually a stream of particles, each one identical to a helium atom which has lost two units of negative electric charge (two electrons, we now know). Because this leaves the stripped helium atoms, also known as alpha particles, with two units of positive charge, they can be manipulated with electric and magnetic fields, steered into beams and accelerated; it's a sign of how fast physics was progressing in the first decade of the twentieth century that by 1909 the Manchester team was using alpha particles produced by natural radioactivity and manipulated in this way to probe the structure of matter.

At that time, adherents to the atomic theory thought of atoms as balls of positively charged stuff with negatively charged electrons embedded in the balls, like pips embedded in a watermelon or plums in a plum pudding (this model had been developed by

J.J. Thomson, who had been Rutherford's mentor in his early days in Cambridge, and who is credited with the discovery of the electron). Rutherford and Geiger had been firing alpha particles through thin sheets of gold foil and monitoring how they were deflected on their journey. The alpha particles that had passed through the foil were detected on the other side using a scintillation screen, where they made little flashes of light.* Geiger had a promising student, Marsden, whom he wanted to encourage, and Rutherford suggested that he could look to see if any of the alpha particles were being reflected by the foil. Nobody expected that he would see much, if anything at all. It was the kind of boring and probably pointless job that is given to a student to provide experience in running an experiment. But to his surprise Marsden saw flashes on the detector screen at a rate of more than one a second. Many alpha particles were being reflected in some way, either deflected through a large angle to one side or bounced back nearly the way they had come. As Rutherford later remarked: 'It was as if you fired a 15-inch shell at a piece of tissue paper and it came back and hit you.' But there was no sudden flash of insight into what was going on.

Rutherford's first thought was that there might be a concentration of negative electric charge deep inside Thomson's 'plum pudding'. This would attract the positively charged and fast-moving alpha particles and send them swinging around the negative charge and back in the direction they came from, like a comet being attracted by the gravity of the Sun and swinging round it before heading back into deep space. Then, after a continuing series of

---

* This is not the famous Geiger counter, but it is the same Geiger.

Ernest Rutherford
Library of Congress/Science Photo Library

careful experiments to build up a clearer picture of what was going on, he hit on a better idea, which fitted the pattern of scintillations more closely. There must be a concentration of *positive* charge at the centre of the atom (now called the nucleus) surrounded by a much larger cloud of negative charge associated with the electrons. Most alpha particles brushed through the electron cloud and went on their way, but the relatively small number that scored a direct hit on the nucleus were reflected by its positive charge and bounced back. Using the statistics of the experiment, where one in a few thousand alpha particles were affected in this way, in 1911 Rutherford concluded that the nucleus was less than one-hundred-thousandth of the size of the atom. The discovery of the concentration of charge at the heart of the atom was announced at a scientific meeting in Manchester, and published in May 1911, although Rutherford only came down firmly in favour of a *positively* charged nucleus in 1912. The explanation of why the negatively charged electrons didn't all fall into the positively charged nucleus had to await the development of quantum theory, but from that moment on there was no doubt that the atom was mostly empty space, and it soon became clear that alpha particles are, in fact, helium nuclei.

Modern measurements have shown that the diameter of the nucleus is in the range of roughly 1.7 femtometres (1.7×10^{-15}m) for hydrogen (the lightest element) to about 12 fm for uranium, the heaviest naturally occurring element. The diameters of atoms range from 0.1 to 0.5 nanometres (1×10^{-10}m to 5×10^{-10}m), so the relative sizes of atoms and nuclei are very much in line with Rutherford's early estimates.

For people unfamiliar with such small numbers, the emptiness of the atom can be pictured more graphically. If the nucleus were the size of a grain of sand, an atom would be the size of the Albert Hall. Very roughly, the difference in size between an atom and a nucleus is also comparable to the size of yourself compared with one of your cells. And for sports fans, if the nucleus were the size of a golf ball, an atom would be about 2.5 kilometres in diameter. You get the picture. It is only electric forces operating in the clouds of almost empty space surrounding tiny nuclei that make it possible for atoms to cling together to make 'solid' objects. It is also the behaviour of electrons in the clouds surrounding nuclei that makes it possible for us to work out what the stars are made of.

The Stars Are Suns and
We Know What They Are Made Of

In 1835 the philosopher Auguste Comte wrote that 'there is no conceivable means by which we shall one day determine the chemical composition of the stars'. In 1859, a technique for determining the chemical composition of the stars was presented in a paper to the Prussian Academy of Sciences. The juxtaposition highlights what an astonishing surprise this was, although that 1859 presentation was far from being the end of the story.

The story had actually begun, unknown to Comte, in 1802, when the English physician and physicist William Hyde Wollaston was studying the spectrum made by sunlight when it is spread out by a triangular glass prism to make a rainbow pattern. He noticed that the pattern was broken up by dark lines, two in the red part of the spectrum, three in the green region, and four at the blue end. He thought these were just gaps between the colours, and didn't follow his discovery up. In 1814, the German industrial scientist Josef von Fraunhofer independently made the same discovery when he was carrying out experiments to improve the quality of the glass used in lenses and prisms. He first noticed the opposite effect from Wollaston – when the light from a flame was passed through a prism, there were two bright yellow lines in the spectrum, at very well-defined wavelengths. He used this pure yellow light to test the optical properties of different kinds of glass, and then looked at the way the glass affected sunlight. At that point, he saw the dark lines discovered, unknown to him, by Wollaston. Because he had bet-ter equipment and high-quality glass, Fraunhofer saw many more lines in the solar spectrum, eventually counting 576 of them, each at a specific wavelength; the overall effect is rather like a barcode. Significantly, he saw the same sort of pattern of lines in the light from Venus and from stars. He never found out what caused the pattern, but to this day they are known as Fraunhofer lines.

The next big step was taken by Robert Bunsen and Gustav Kirchhoff, working in Heidelberg in the 1850s. They knew, as all chemists at the time did, that different substances sprinkled into a clear flame would make it flare up with different colours – yellow for a trace of sodium (as in common salt, sodium chloride), blue-green for copper, and so on. They had a very good type of burner to use in these 'flame tests', the one named after Bunsen himself,* and built an apparatus incorporating a prism and an eyepiece like that of a microscope to study the light in detail (this was the first spectroscope). When they analysed the coloured light produced in this way using spectroscopy, they found that in the heat of the flame

---

* The basic burner was designed by Michael Faraday and improved by Peter Desaga, Bunsen's assistant, who marketed it under Bunsen's name.

each element produces distinctive sharp lines at specific wavelengths. The two yellow lines noticed by Fraunhofer are produced by sodium, copper makes sharp lines in the blue-green part of the spectrum, and so on. They realised that any hot object produces its own pattern of distinctive lines in the spectrum.* Then serendipity stepped in.

One evening, while they were working in their laboratory in Heidelberg a major fire broke out in Mannheim, about ten miles away. They were in the right place at the right time to analyse the light from the fire using spectroscopy, and were able to identify lines corresponding to strontium and barium in the spectrum. According to a story they repeated in different versions at different times later, a few days after the fire Bunsen and Kirchhoff were walking along the River Neckar when Bunsen said something like: 'If we can determine the nature of substances burning in Mannheim, we should be able to do the same thing for the Sun. But people would say we are mad to dream of such a thing.'

Back in the lab, they tested the mad idea. Kirchhoff almost immediately identified the familiar double lines of sodium in the yellow part of the solar spectrum, then they found, with Kirchhoff taking the lead, that many of the dark Fraunhofer lines occurred at wavelengths where specific elements produce bright lines when heated in the flame of a Bunsen burner. The implication is that although hot things produce bright lines in the spectrum, when light passes through cool things they absorb light at the corresponding

---

* An explanation of *why* this happens had to await the development of quantum theory in the twentieth century; but that didn't matter to the chemists of the time.

wavelengths, making dark lines. Light from the hot interior of the Sun must be passing through cooler outer layers to produce this effect. It *was* possible to determine what the Sun was made of. Kirchhoff was so astonished that he exclaimed, referring back to their riverside conversation, 'Bunsen, I have gone mad!' Bunsen replied, 'So have I, Kirchhoff.' It was this work that formed the basis of Kirchhoff's presentation to the Prussian Academy on 27 October 1859. It really was possible to say what the Sun and stars were made of. Or was it?

At first, everything looked good. The great triumph of the new technique for analysing light from stars came following an eclipse of the Sun visible from India on 18 August 1868 – the first eclipse following Kirchhoff's realisation that Fraunhofer lines are caused by specific elements blocking light from the Sun at particular wavelengths. During an eclipse, with the light from the main disc of the Sun blocked out by the Moon, it is possible to study the fainter light from regions just above the surface. The French astronomer Pierre Janssen did just that, and found a very bright yellow line close to the expected sodium lines. This feature was so bright that he realised he could still study it even after the eclipse, and he made more observations before returning to France. Meanwhile, an English astronomer, Norman Lockyer, had developed a new spectroscope that he used to study light from the outer regions of the Sun on 20 October 1868. He found the same yellow line. Janssen and Lockyer were both credited with the discovery. But it was Lockyer alone who took the bold step of claiming that the line must be produced by an element that was unknown on Earth, and gave it the name helium, from the Greek word for the Sun. The suggestion remained controversial until 1895, when William Ramsay found that a gas released by uranium (we now know, as a result of radio-active decay) produces the same bright yellow line in the spectrum. An element had actually been 'found' in the Sun 27 years before it was found on Earth. At the beginning of the twentieth century, the plethora of elements identified by spectroscopy seemed to be tell-ing us that although the Sun was being kept hot by some unknown process, its composition was very much like that of the Earth. But this interpretation of the evidence was wrong. There were still sur-prises in store.

Although the interpretation was wrong, it was based on what seemed to be solid evidence. At the end of the nineteenth century, the state of knowledge about the Sun's composition was summed up by Henry Rowland in a series of tables identifying 36 elements and giving details of the strength of their spectral lines. This infor-mation revealed the relative proportions of these elements – how many atoms of oxygen for each atom of carbon, and so on – which matched the proportions seen on Earth. Largely as a result of Rowland's work, the idea that the Sun's composition was much the same as that of the Earth persisted for a quarter of a century. Then came the first surprise.

In 1924 Cecilia Payne was working for a PhD at Harvard University. Payne was English, and had studied at Newnham College in Cambridge, but had not been allowed to take a degree there, let alone a PhD, because she was a woman. That was why she had moved to America, where in 1925 she would be the first woman to be awarded a PhD by Radcliffe College, based on her work at Harvard College Observatory. This was just the beginning of a glittering career,* but nothing surpassed what she achieved in the mid-1920s. Her starting point was recent work by the Indian physicist Meghnad Saha which had explained how details of the Fraunhofer lines were affected by the physical conditions (temperature, pressure and so on) in different parts of a star. Armed with this information, she was able to work out the proportions of eighteen elements in several stars more accurately than anyone before her, showing that they were essentially the same for all stars once allowance was made for these effects. Most of these abundances were in line with Rowland's tables for the solar abundances. But there was one dramatic difference. According to Payne's calculations, there was overwhelmingly more hydrogen and helium in the stars than everything else put together.

When Payne prepared the draft of her thesis, including this discovery, her supervisor, Harlow Shapley, sent it to be reviewed by Henry Norris Russell, a senior astronomer at Princeton. He said that this result was 'clearly impossible'. Back in 1914, in an article on 'The Solar Spectrum and the Earth's Crust', Russell had written:

> The agreement of the solar and terrestrial lists is such as to confirm very strongly Rowland's opinion that, if the Earth's crust should be raised to the temperature of the Sun's atmosphere, it would give a very similar absorption spectrum. The spectra of the Sun and other stars were similar, so it appeared that the relative abundance of elements in the universe was like that in Earth's crust.[4]

---

* Mostly carried out under her married name, Cecilia Payne-Gaposchkin.

Cecilia Payne-Gaposchkin
Smithsonian Institution/Science Photo Library

And he still held to that view. On Shapley's advice, when Payne formally submitted her thesis in 1925 she included the sentence 'the enormous abundances derived for [hydrogen and helium] in the stellar atmospheres is almost certainly not real'.

But this was an idea whose time had come. In 1928 the astronomer Albrecht Unsöld, working in Göttingen, made a study of the solar spectrum, and came to the conclusion that the atmosphere of the Sun must be mostly composed of hydrogen. A young Irish research student, William McCrea, was visiting Göttingen at the time, and developed this suggestion with a calculation that showed there were a million times more hydrogen atoms in the solar atmosphere than the amount of everything else, except helium, put together.* His PhD, for a thesis on 'Problems Concerning the Outer Layers of the Sun', was awarded by Cambridge University in 1929. About the same time, Russell was changing his mind about the impossibility of Payne's results. Building from Unsöld's work, and also using the Saha equations, Russell carried out a detailed study of the solar spectrum, which provided relative abundances for 56 elements. This was the best set of data yet on the Sun's composition, including evidence that 'the great abundance of hydrogen can hardly be doubted', even though he described it as 'almost incredibly great'. Russell was careful to give full credit to Payne when he published his own work, but because he was already an established scientist his paper made a big splash at the time, and he often received her share of the credit for the discovery. His work did go

* Much later, McCrea was my supervisor when I studied astronomy at the University of Sussex, but his brilliance did not rub off on me!

further than hers, but hers did come first; in 1962, the astronomer Otto Struve described her thesis as 'the most brilliant PhD thesis ever written in astronomy' up to that time. What had been 'impossible' in 1925 was merely 'almost' incredible in 1929.

But there was still more to surprise astronomers, hinted at in Russell's comment in 1914 that 'it appeared that the relative abundance of elements in the universe was like that in Earth's crust'. If the stars were not made of the same elements in the same abundances as the Earth's crust, then the composition of the Universe is not like that of the Earth's crust. Specifically, the Universe must contain a lot more hydrogen and helium. Just how much more only became clear nearly three decades after the pioneering work of Payne, Unsöld, McCrea and Russell.

By the end of the 1920s, astronomers had a surprisingly good understanding of the nature of a star like the Sun, even though they did not know the exact details of how it generated heat in its interior. A star is basically a ball of hot gas which is balancing two opposing forces to maintain equilibrium. Gravity is trying to pull the ball together and make it shrink, while the pressure generated by the heat in its interior is pushing outwards to hold it up. Astronomers can calculate the mass of the Sun by studying the orbits of the planets, held in place by the Sun's gravity, so they know how strong the inward force is. Because of the equilibrium, this means they know how strong the outward force is, which tells them about the conditions inside the Sun, including the temperature at its core. The details were worked out by the pioneering astrophysicist Arthur Eddington, and published in a book, *The Internal Constitution of the Stars*, in 1926. By then, thanks to Albert Einstein, physicists knew that energy could be released by nuclear fusion. A great deal of energy could be released in this way if four hydrogen nuclei (single protons) could be converted into one helium nucleus (alpha particles, each composed of two protons and two neutrons bound together), because each helium nucleus has less mass than the combined mass of four individual protons. The energy released in each such fusion is equal to this 'lost' mass multiplied by the square of the speed of light. Even before astronomers realised just how much hydrogen the Sun and stars contain, all potentially available to be involved in this process, Eddington proposed that their heat is generated in this way since 'the helium which we handle must have been put together at some time and some place'. The question was, how?

The search for an answer to the question was handicapped by an unfortunate coincidence. In the 1930s, astrophysicists developed more detailed 'models' (sets of equations describing what was going on) of stellar interiors. They found that the pressure which holds a star up has two parts. One is the regular process we think of as pressure, with particles bouncing around and colliding with one another, like the molecules of air in a balloon. But the interior of a star is so hot that negatively charged electrons are stripped from positively charged nuclei. The resulting sea of charged particles interacts with the electromagnetic radiation – light, X-rays and so on – released at the heart of the star and making its way to the surface. This produces an additional outward force, known as radiation pressure. A star like the Sun is stable when the combination of both these kinds of pressure balances the inward tug of gravity. But it turns out there are two ways to achieve this balance.

The conventional pressure depends on the number of particles there are. Electrons are so much lighter than protons and neutrons that they can be ignored for this purpose, so what matters is the number of atomic nuclei. But the radiation pressure does depend on the number of electrons. A hydrogen atom only has one electron, so it can only contribute one electron per nucleus, but a helium atom has two electrons, so it can contribute two electrons per nucleus, and so on. So the proportions of the overall pressure contributed by conventional pressure and by radiation pressure depend on how many nuclei of heavy elements there are in the mix. The unfortunate coincidence is that for a star with the mass and brightness of the Sun, or any similar star, the inward tug of gravity can be balanced by a combination of the two pressures *either* if at least 95 per cent of the star is composed of a mixture of hydrogen and helium, *or* if there is just 35 per cent light stuff and 65 per cent heavier elements. In the 1930s, having only just realised that stars are not made entirely of heavier elements, astronomers leaned towards the second option. To accept that elements like those we find on Earth made up no more than 5 per cent of the Sun and stars was too big a leap for them to accept.

So the first attempts to explain how energy is generated in stars – how hydrogen nuclei are combined to make helium nuclei – had the assumption of 35 per cent hydrogen built in to them. This misunderstanding affected the work of the first people to attempt to explain the process, initially carried out by Fritz Houtermans and Robert Atkinson in collaboration and then developed by Atkinson. The essence of this idea is that heavier nuclei absorb four protons one after another, and then spit out alpha particles – helium nuclei. It turns out that this process is important in some stars a bit more massive than the Sun, but the process which actually releases energy inside the Sun is much simpler. This starts with two protons getting together and spitting out a positron (a positively charged counterpart to the electron) to make a deuteron, a nucleus consisting of a single proton and a single neutron bound together. The addition of another proton makes a nucleus of helium-3, and when two helium-3 nuclei interact they form a nucleus of helium-4 (two protons and two neutrons, an alpha particle) with two protons being ejected. The net effect is that four hydrogen nuclei have been converted into one helium nucleus, and energy has been released. The essence of this 'proton-proton chain' was worked out by Charles Critchfield in 1938, but it was only fully understood, with the implication that the Sun is at least 95 per cent made of hydrogen and helium, in the 1950s.

You need a lot of hydrogen to make this work, because the chances of any two hydrogen nuclei bumping into one another with enough force to make a deuteron, even under the extreme conditions at the heart of the Sun, are small. Modern calculations, greatly aided by the advent of high-speed electronic computers, tell us that it would take an individual proton, bouncing around in the heart of the Sun where the temperature is about 15 million degrees Celsius, 14 billion years before it was involved in a head-on collision with a partner to form a deuteron. Some take longer, some take less time, but statistically speaking just one collision in every 10 billion trillion will trigger the start of the proton-proton chain. And the other steps in the chain are comparably unlikely. Each time four protons are converted into a single helium nucleus, just 0.7 per cent of the mass is converted into energy. Yet in spite of the rarity of these events and the small amount of mass-energy released each time, overall the Sun is converting 5 million tonnes of mass (the equivalent of a million medium-sized African elephants) into energy every second, as 616 million tonnes of hydrogen is converted into 611 million tonnes of helium. It has been doing this for 4.5 billion years, but it started out with so much hydrogen that so far it has processed only about 4 per cent of its hydrogen fuel into helium ash.

As far as the composition of the Sun (and similar stars) is concerned, the situation is even more extreme than the simple calculations carried out in the 1930s suggested. They said that *at least* 95 per cent of the Sun must be in the form of hydrogen and helium. We now know, from a combination of observations and computer modelling, that in terms of mass the Sun is made up of some 71 per cent hydrogen, roughly 27 per cent helium, and less than 2 per cent of everything else put together. In terms of the number of atoms (nuclei) the numbers are even more impressive. Hydrogen nuclei make up 91.2 per cent of the Sun, helium 8.7 per cent, and everything else just 0.1 per cent. These numbers apply to the proportions of the chemical elements in stars across the Universe, and planets are insignificant (in cosmic terms) specks of dust compared with their parent stars (the Sun is as big as 1.3 million Earth-sized planets put together). Everything that matters to us is part of the 2 per cent, or 0.1 per cent if you are counting atoms, a kind of afterthought of creation. This was one of the biggest surprises of science. Yet, arguably even more surprisingly, that 2 per cent has produced life, including ourselves. How it has done so forms one of the other pillars of science.

There is No Life Force

The idea that life is special – that living things are powered by a mysterious 'life force' that inanimate objects lack – is older than history, and was discussed by the philosophers of ancient Egypt and Greece. It seems like common sense. But as so often with our understanding of the world, common sense is a bad guide to reality.

The beginning of a proper understanding of how living things work came with experiments carried out by the French chemist Antoine Lavoisier and his colleague Pierre Laplace in the 1780s. They put a guinea pig in a container which stood inside another container packed with ice, and measured how much ice was melted by the animal's body heat in a certain time. They also measured how much 'fixed air' (now known as carbon dioxide) the animal breathed out. They found that this was the same as the amount produced by burning charcoal to melt the same amount of ice. Animals were seen to obey the same laws as burning charcoal, or candles.

In the following decade, though, another discovery seemed to suggest at first that there really is a life force. Famously, the Italian physician Luigi Galvani accidentally discovered that a pair of frog's legs that had been dissected twitched when they were in contact with iron. The story is a little more complicated than many popular accounts suggest, though, and it is worth going over the details to see how a scientist's mind works.

Galvani carried out many kinds of experiments, and in his lab there was a hand-cranked machine which generated static electricity through friction, like the shock you can get when touching a metal object after walking across some kinds of carpet. One day he was dissecting a pair of frog's legs, using a scalpel which had touched the machine and picked up an electric charge. When the scalpel touched the sciatic nerve of one of the legs, the leg kicked as if it were still alive. This led him to experiments which showed that legs from a dead frog would twitch if they were connected directly to the electrical machine, or if they were laid out on a metal sheet during a thunderstorm, when there was lightning in the air. But his key observation came about by accident. When he was getting the sets of legs ready for his experiments, Galvani would hang them up to dry on brass hooks in the open air. One of these hooks touched an iron fence, and the legs kicked, although there was no outside source of electricity. When he took the legs and hook inside, keeping them well away from his electrical machine, and touched the hook onto iron, the legs twitched again. It happened every time, with every set of legs.

He believed that this proved the presence of some kind of 'animal electricity', different from the 'static' electricity that makes lightning, or that we can make by friction. This animal electricity was supposed to be a kind of fluid, manufactured in the brain, which was carried to the muscles by nerves, and stored there until needed. Galvani's compatriot, Alessandro Volta, disagreed. He said that the electricity that produced the twitching was not something to do with a life force, but the result of an interaction involving the metals the dissected legs were touching. This led him, through a series of experiments, to invent a device for making electricity. It was a pile of alternating silver and zinc discs, separated by cardboard discs soaked in brine. When the top of the pile was connected to the bottom by a wire, an electric current flowed in the wire. The 'Voltaic pile' was the first electric battery.

Volta's invention was developed at the Royal Institution in London and applied there by Humphry Davy in dramatic experiments that used electricity to break down compounds into their constituent parts, revealing the existence of 'new' metals, including potassium and sodium. But instead of quashing the idea of a life force, Davy's experiments encouraged some supporters of the idea. In particular, a London-based surgeon, John Abernethy, saw a link between electricity and something he called Vitality – essentially his name for the life force. His conclusions were attacked by one of his colleagues, William Lawrence, triggering a debate that raged in the second decade of the nineteenth century (it is no coincidence that Mary Shelley wrote her novel *Frankenstein* just at this time; Lawrence was Percy Bysshe Shelley's doctor from 1815 to 1818). The study of electricity could not settle the issue. But the surprising result of a chemical experiment carried out in 1828 should have laid vitalism to rest.

By the end of the eighteenth century, chemists were beginning to get a handle on how different substances combined to form more complicated compounds. It was soon clear that carbon can indeed form a great variety of complicated combinations with other things, and that living things are largely made up of such complicated carbon compounds. The chemistry of such carbon compounds became known as organic chemistry, regarded as something distinct from the chemistry of ordinary 'inorganic' things like water, and was associated with the idea of vitalism. Organic compounds, it was thought, could only be manufactured by living things, thanks to the power of the life force.

In 1773, Hilaire Rouelle, a French chemist, had isolated crystals of a previously unknown substance from the urine of various animals, including people. This became known as urea, and it was something of a puzzle because even at the time it was clear that it is a relatively simple compound (its modern formula is $H_2N\text{-}CO\text{-}NH_2$). It didn't really seem complicated enough to require the influence of a life force in its manufacture. As it turned out, it wasn't.

In 1828, the German chemist Friedrich Wöhler was trying to make ammonium cyanate by reacting cyanic acid with ammonia. The stuff produced by his experiment, however, was not ammonium cyanate. Careful analysis showed him that it was urea, identical to the stuff extracted from urine. His surprise was expressed in the introduction to the paper he wrote reporting the discovery, in 1828:

This investigation gave the unexpected result that by combining cyanic acid and ammonia urea is generated. Quite a peculiar fact in that it represents the artificial (in vitro) formation of an organic compound, so-called 'animalischem Stoff', out of inorganic compounds.

He was less formal in a letter he wrote that year to a colleague, Jacob Berzelius, to inform him that he [Wöhler]:

> is capable of producing urea requiring neither kidneys nor any animal, may it be man or dog. Ammonium cyanate is urea ... it is by no chemical means different from urea of the urine, which I have produced all by myself.

This might have been something of a killer blow to the idea of a life force. But one reason why it did not have the immediate impact that we might expect with hindsight is revealed in the quote from that letter. Wöhler's tests showed that chemically speaking, urea and ammonium cyanate are identical. A molecule of ammonium cyanate does actually contain the same atoms as a molecule of urea, but in a different geometrical arrangement. Such non-identical twin molecules are now known as isomers, and Wöhler was much more interested in following up the discovery of isomerism than in getting involved in the vitalism debate. Besides, urea is a relatively simple substance, and supporters of the idea of a special kind of chemistry of life could (and did) argue that it hardly counted as an organic molecule at all. There were many other organic molecules that were more complicated and could not be synthesised.

The only way to lay vitalism to rest was to synthesise a lot more of these complicated organic molecules, starting from simple inorganic molecules – a process known as 'total synthesis'. Wöhler's discovery had been a happy accident, a stroke of serendipity. But in 1845 another German, Adolph Kolbe, deliberately set out to make organic compounds from inorganic substances. He gave himself the task of converting carbon disulphide, an inorganic compound easily made out of its constituent parts, into acetic acid, or vinegar, an organic compound produced naturally by fermentation. Kolbe's success was the second complete synthesis of an organic compound from inorganic precursors, without involving any biological processes. But even two examples left a lot of organic molecules to be investigated.

In the 1850s, the Parisian Marcellin Berthelot, an almost evangelical opponent of vitalism, set out to use total synthesis to manufacture every organic molecule known at the time. His conviction was that all chemical processes are based on the action of physical forces which can be studied and measured like the forces involved in mechanical processes. His programme of total synthesis of every organic substance followed logically from that conviction; it was an impossible dream, but he did enough to show that he was right.

Berthelot planned a step-by-step approach. He started out with simple compounds containing carbon and hydrogen (hydrocarbons, such as methane), converted them to alcohols (which contain an OH group, essentially water with one hydrogen atom missing, so it can link to other things), then changed them into esters (where the OH group is replaced by a more complex 'alkoxy' group), converted them into organic acids (which contain even more complex groups) and so on. Berthelot had many successes. He was able to make formic acid (the chemical ants use to sting with) using the step-by-step approach just outlined, acetylene (which he named) by sparking an electric arc between carbon electrodes in an atmosphere of hydrogen, and benzene by heating acetylene in a glass tube.

Marcellin Berthelot
Science Photo Library

Benzene was a crucial step. Each molecule of benzene is built upon six carbon atoms joined in a ring. Benzene is found naturally in crude oil, which is the remains of living organisms, and these ring molecules, which are integral components in a huge variety of compounds, are particularly important in the chemistry of life. The branch of chemistry involving the reactions of such ring molecules is now known as aromatics.

Berthelot's epic programme of total synthesis was far too ambitious for one man to complete, but he established something which is now one of the pillars of science. He showed that it is possible to manufacture organic substances from four elements that are found in all living things – carbon, hydrogen, oxygen and nitrogen. These are so important, and so often (always!) found together in organic material, that they are collectively referred to as CHON. Berthelot's opus on the synthesis of organic chemicals, *Chimie organique fondée sur la synthèse*, was published in 1860. It should have sounded the death knell of vitalism. But the idea that living things, including ourselves, are nothing more than collections of carbon compounds which operate through the action of physical forces like the forces involved in mechanical processes was so hard to stomach – so counter to 'common sense' – that even at the end of the nineteenth century the idea that there was something special about the chemistry of life, and that some 'life force' was involved in what were called vitalistic processes, was still a subject for debate. As respectable a scientist as Louis Pasteur argued in its favour. The final refutation of this idea came in 1897, from the work of the German Eduard Buchner.

One of the last puzzles that gave ammunition for the vitalists was fermentation. Fermentation converts foods such as sugar into simpler compounds such as alcohol and carbon dioxide, and releases energy which powers living cells. But did it always involve living cells? Buchner tackled the question using alcohol production, which involves yeast, a living organism. Yeast is essential for this process; but Buchner wanted to test whether this was because the yeast cells were alive, or because they contained some chemical substance (a catalyst) which encouraged the conversion of sugar into alcohol and carbon dioxide by inorganic reactions.

Buchner started with living yeast cells, then subjected them to a series of indignities which killed them and reduced them to their constituent chemical parts. Dry yeast cells were mixed with quartz sand and a soft crumbly rock, then ground up with a pestle and mortar. The mixture became damp as the yeast cells burst and released their contents. The damp mixture was then squeezed to extract a 'press juice' used in the experiments.

When sugar solution was mixed in to the freshly pressed yeast juice, bubbles of gas were produced, eventually covering the liquid with froth. Chemical tests showed that carbon dioxide and alcohol were being produced in exactly the same proportions as in fermentation with live yeast. But there were no living yeast cells in the extract.

Following up this work, Buchner found that the key chemical substance involved is an enzyme, which he called zymase. Zymase is manufactured inside yeast cells, so in that sense life is involved in the process of fermentation, but the key point is that zymase itself is an inanimate chemical substance, and fermentation occurs whether the yeast is alive or dead. Enzymes are crucial players in many biological processes, but it is now possible to synthesise enzymes chemically without biology being involved. As Buchner later put it:

> The difference between enzymes and micro-organisms is clearly revealed when the latter are represented as the producers of the former, which we must conceive as complicated but inanimate chemical substances.

Zymase is, indeed, one of the enzymes that it is now possible to synthesise without biology being involved. But the essential point, worth reiterating, is that the chemistry carries on whether the yeast is alive or dead. In January 1897 Buchner sent his key scientific paper, *Alkoholische Gährung ohne Hefezellen* (*On alcoholic fermentation without yeast cells*), to the journal *Berichte der Deutschen Chemischen Gesellschaft*.

Buchner was awarded the Nobel Prize for chemistry in 1907, 'for his biochemical researches and his discovery of cell-free fermentation'. This is as good a date as any to choose to mark the death of vitalism. But this left another question. If there is no life force, how did life originate?

Charles Darwin thought hard about this puzzle, and speculated that life might have got started in a 'warm little pond' on Earth laced with the right chemical ingredients. But he realised that this could not happen today. As he wrote to Joseph Hooker in 1871:

> It is often said that all the conditions for the first production of a living organism are now present, which could ever have been

present.—But if (& oh what a big if) we could conceive in some warm little pond with all sorts of ammonia & phosphoric salts,— light, heat, electricity &c present, that a protein compound was chemically formed, ready to undergo still more complex changes, at the present day such matter w^d be instantly devoured, or absorbed, which would not have been the case before living creatures were formed.

Half a century after Darwin wrote those words, Alexander Oparin, a Russian biochemist who had been born in 1894 and graduated from Moscow State University in 1917, the year of the Russian Revolution, put this kind of speculation onto a proper scientific footing. It was actually in 1922, at a meeting of the Russian Botanical Society, that he first aired his ideas, which he developed into a book, *The Origin of Life*, published in 1924. What stimulated his thinking was the recent discovery (thanks to spectroscopy; none of the scientific pillars stand alone!) that the atmospheres of Jupiter and the other giant planets of our Solar System contain a great amount of gases such as methane, the kind of thing Darwin (among others) envisaged as the feedstock for life. The atmosphere of the Earth today is rich in oxygen, which is very reactive. It is produced by life, but if it were not being constantly replenished it would all get used up in forest fires, weathering of rocks, and other processes. Oparin suggested that in order for life to have got started in some warm little pond, when the Earth was young its atmosphere must have been like those of the giant planets. Such a 'reducing' atmosphere might have contained methane, ammonia, water vapour, and hydrogen, and might build up organic molecules step by step, as in Berthelot's experiments. But it would have had no oxygen, which would have reacted with and destroyed the precursors of life.

Oparin's argument has been summed up in a few steps:

- There is no fundamental difference between living things and lifeless matter. The complexities of life must have developed in the process of the evolution of matter.
- The infant Earth possessed a strongly reducing atmosphere, containing methane, ammonia, hydrogen and water vapour, which were the raw materials for the evolution of life.
- As molecules got larger and more complex, their behaviour also became more complex, with interactions between molecules determined by the shape of the molecules and the way they fitted together.
- Even at this early stage, development of new structures was governed by competition, a struggle for existence as complex structures 'fed' off simpler molecules, and Darwinian natural selection.
- Living organisms are open systems, which take in energy and raw materials from outside, so they are not restricted by the second law of thermodynamics.

The last point is an important one which is often overlooked. The second law of thermodynamics is famous as the law of nature which tells us that things wear out and the amount of disorder in the world (measured in terms of a quantity called entropy) always increases. The standard example is a drinking glass that falls off a table and shatters on the floor. The shattered glass is more disordered than the original drinking vessel; entropy has increased. And you never see fragments of broken glass on the floor spontaneously rearrange themselves into a drinking vessel – negative entropy. But life seems to get around this law. Somehow, life creates order out of chaos, reversing the flow of entropy. But it can only do so on a local basis. Just as making a drinking glass requires an input of energy, so making living things and maintaining life requires an input of energy. For animals like ourselves, this comes from our food – ultimately from plants, since even if we eat meat, the meat comes from animals that ate plants. For plants the energy comes ultimately from sunlight. The living Earth is like a bubble of reversed entropy flow, all feeding off the energy stream flowing out from the Sun. And that is more than compensated for by the vast increase in entropy that is associated with the processes that keep the Sun shining.

Oparin's specific suggestion was that with the aid of energy from sunlight or some other outside source, such as lightning, in the kind of reducing atmosphere he envisaged, entropy could have 'run backwards' to build up complex molecules containing carbon – organic molecules. Such things could grow into sheets and tiny droplets or even little hollow bubbles, the sort of thing that might develop into cells. Oparin's work went largely unnoticed outside his homeland at the time, but as if to show that the time was indeed ripe for it, the British researcher J.B.S. Haldane independently came up with what was essentially the same idea in 1929. And it was Haldane (who, as we saw earlier, had a gift for memorable scientific quips) who thought up a catchy name for the hypothetical 'warm little pond' in which all of this took place – the primordial soup. The next step was to try to create, or recreate, the conditions that existed in the primordial soup in the laboratory. But although those experiments were successful up to a point, they raised new questions, and were overtaken by another surprise, which has itself become one of the pillars of science.

The Milky Way is a Warehouse Stocked with the Raw Ingredients of Life

Two decades after Oparin suggested that life might have originated under a reducing atmosphere on the early Earth, Harold Urey, a chemistry professor at the University of Chicago, gave his students a lecture about what he referred to as 'the Oparin-Haldane hypothesis'. One of those students, a recent graduate named Stanley Miller, was sufficiently intrigued to ask if he could work for his doctorate by building an experiment to test the idea – a 'warm little pond' in miniature sealed within the glass vessels of a laboratory and containing the mixture of materials suggested by Oparin and Haldane.

Urey agreed to supervise the work, and the result became famous as the Miller–Urey experiment.

The centrepiece of the experiment was a 5-litre glass flask which held a mixture of methane, ammonia, water vapour, and hydrogen. The water vapour was provided continuously from a second half-litre flask of boiling water linked to the main flask by tubes; the vapour passed through the main flask and then condensed, with the hot gases from the larger flask continuing through a cooling chamber, shaped like a U-bend, and going back to the boiling flask to complete the loop. The U-bend provided a trap in which liquid could be caught and drained off through a tap. To provide energy by mimicking the action of lightning, electric sparks flashed through the mixture in the main flask.

In the original form of the experiment, the liquid trapped in the U-bend was drained off and analysed once a week. But it didn't take more than a single week for the experiment to prove a spectacular success, with results well worth the award of a PhD. In less than a day, the liquid in the U-bend had turned pink. When the first week's worth of liquid was drained off and analysed, Miller found that more than 10 per cent of the carbon from the original mixture of gases sealed into the 5-litre flask had been converted into organic compounds. The most important of these were amino acids, complex organic molecules which are themselves the components of proteins, the building blocks of life. There are just twenty amino acids which combine with one another in different ways to form all the proteins in your body. The Miller–Urey experiment had made thirteen of them in just a week. The results were published in the journal *Science* in 1953. They were seen as being just one step short of making life itself, and Miller dedicated his entire scientific career (he died in 2007) to refining his experiment and improving it, with longer and longer runs, in the hope of taking that extra step. He was undaunted when the geologists decided that the Earth had probably not started out with a reducing atmosphere at all. The best evidence is that the early

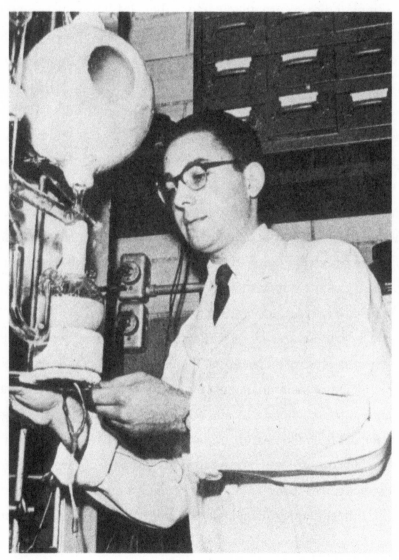

Stanley Miller
Science Photo Library

atmosphere of our planet was composed of the same mixture of gases that spews from volcanoes today – carbon dioxide, nitrogen and sulphur dioxide prominent among them. Miller simply adapted his apparatus to accommodate this mixture of materials and tried again, once again producing a wide variety of complex organic molecules from the simple feedstock. If nothing else, he established that provided there is an input of energy it is not only easy but inevitable that molecules as complex as amino acids will be built up from simple compounds. Ironically, though, he needn't have bothered trying to explain how such molecules arrived on Earth. The big surprise stemming from observations made at the end of the twentieth century and into the early twenty-first century, shaking up our understanding of the origin of life on Earth, is that the chemistry of the warm little ponds on the young Earth may have *started* with compounds like amino acids.

We have moved on a long way since the pioneering work of the nineteenth-century biochemists. If one of their heirs today wanted to manufacture the fundamental molecules of life, proteins and the famous nucleic acids DNA and RNA, he or she wouldn't bother starting out from the mixture of gases that might have been found in a reducing atmosphere, or even from the mixture of gases belched out by volcanoes today.* Much more complex and interesting feedstock, things like formaldehyde and methanol, is available from

---

* To my surprise, after I had written this section, in the autumn of 2019 researchers based at Ludwig Maximilian University of Munich in Germany reported results of recent laboratory experiments in which complex organic molecules were built up from ingredients such as water and nitrogen. Some people are gluttons for punishment.

chemical suppliers, and is probably on the shelves of any decent biochemical lab. Of course, the reason it is so readily available is that someone else has gone to the trouble of manufacturing it by total synthesis, on an industrial scale. The bombshell is that the Universe has done the same thing, on a vastly bigger scale and for a large number of the precursors of life.

The story starts in the 1930s, when simplest molecular compounds of carbon and hydrogen (CH) and carbon and nitrogen (CN) were found in clouds of gas and dust in space (nebulae) using spectroscopy. But the story only began to get interesting in the 1960s, when new technology extended the range of wavelengths that could be investigated in this way. Small molecules, like those first two to be identified in space, produce lines in the visible part of the spectrum. But larger molecules produce equivalent features in the spectrum at longer wavelengths, in the infrared and radio parts of the spectrum. So their identification had to await the development of suitable technology, in the form of infrared and radio telescopes, to make the appropriate identifications. Even then, since nobody expected to find complex molecules in space it took a while for astronomers to realise what it was they were seeing. Then the penny dropped, and they began actively searching for molecules in space, seeking out bigger and more complex varieties in a competition to find the one with the most atoms linked together.

The third molecule found in space was the so-called hydroxyl radical, OH, identified in 1963. But it was the next discovery, made in 1968, that made people begin to sit up and take notice. This was the four-atom molecule ammonia, NH_3. It was the first indication that more than two atoms could get together under the conditions of interstellar space to make molecules. Water (H_2O) was one of the first three-atom molecules to be identified, but much more excitement was stirred by the discovery of formaldehyde (H_2CO), the first organic molecule found in space. A couple of hundred interstellar molecules have now been identified, including urea* and ethyl alcohol. The discovery of ethyl alcohol was particularly interesting, not just because it gave headline writers in popular papers an opportunity to refer to clouds of 'vodka in space' but because each molecule is made up of nine atoms, CH_3CH_2OH. There are a few definitely identified molecules with ten or more atoms each, but one that is particularly intriguing is glycine, H_2NH_2CCOOH. Glycine is an amino acid, one of the building blocks of proteins. What Miller could do in a 5-litre flask in his lab the Universe can do in vast clouds of gas in space.

Another significant discovery is a twelve-atom molecule called iso-propyl cyanide, $(CH_3)_2CHCN$, identified in 2014. This is significant because the notation $(CH_3)_2$ means that two separate CH_3 units branch off from the same carbon atom; this is a structure similar to that of many of the complex molecules of life found on Earth, including some of the amino acids. Two years later, in 2016, astronomers detected the ten-atom molecule propylene oxide, CH_3CHCH_2O in a cloud of gas and dust called Sagittarius B2. The specially interesting feature of this molecule is that it has a property called chirality, which is essentially handedness. It comes naturally in left-handed and right-handed varieties – but only one kind was

---

* Detected in 2014 and definitely manufactured in interstellar clouds without the use of kidneys, either man or dog.

found in 2016. Helices have chirality – they can twist either to the left or to the right. Life on Earth is neatly divided into both kinds of chirality. Amino acids are almost entirely left-handed, while the helices of RNA and DNA are right-handed. The propylene oxide seen in clouds like Sagittarius B2 will have their chirality determined by the action of light* from stars which only allows one kind of handedness to be imprinted on the molecules within an individual cloud of gas and dust, although the observations aren't yet detailed enough to pick out which handedness dominates in this case. Since stars and planets form from such clouds, the implication is that the handedness is imprinted on the components of life before they even reach the surface of a planet. All the planetary systems that form from the same cloud should have the same handedness. But how exactly do these molecules form in space, and how can they reach the surface of a planet?

Mention of 'dust' in interstellar clouds might not conjure up quite the right image in your mind. Studies of the electromagnetic radiation from these clouds across a wide range of wavelengths show that the dust is made up of tiny particles, like the particles in cigarette smoke. The particles are made of things like carbon and the oxides of silicon, and they are covered in ices made of frozen ammonia and methane, as well as familiar water ice. If two atoms, or two small molecules, or one large molecule and one small molecule, collide in space, they will most probably bounce off one another or be broken apart by the collision. But the icy surfaces of dust grains provide places where atoms and molecules can stick.

---

* For the technically minded, circularly polarised light.

When simple substances stick onto the ice they have an opportunity to join together to make more complex substances which later escape from the ice, perhaps when the grains themselves are involved in collisions, or as a result of the impact of cosmic rays, fast-moving particles ejected by stellar activity. These ideas have been tested in the lab, where icy particles like the ones that exist in space have been cooled to −263°C to mimic the cold of space and bathed in ultraviolet light to mimic the energy provided by stars. Chemical reactions take place on the surfaces of the grains in just the way I have described.

This is not a very quick process. It takes a long time to build up molecules as complex as glycine or iso-propyl cyanide. But there has been plenty of time. The Universe is about 13.8 billion years old, and our Milky Way galaxy is only a little younger. Even the Solar System and the Earth itself are around 4.5 billion years old. Fossil remains show that single-celled life forms existed on Earth at least 3.8 billion years ago, and it is a puzzle how things like carbon dioxide, water and sulphur dioxide could produce such life forms in such a short time, starting from scratch. But it is much less of a puzzle how such feedstock, plus methane and ammonia, could produce things like glycine or iso-propyl cyanide in ten or more billion years, more than twice the present age of the Earth.

How much complex organic material might there be out there? Our Milky Way contains several hundred million stars more or less like our Sun, and a variety of astronomical observations suggest that the mass of all the gas and dust between the stars is about 10 per cent of the mass of all the stars. That means at least 10 million times the mass of the Sun. And we can see what happens when such clouds of gas and dust are pulled together by gravity to make new stars and planets.

There is a system known as IRS 46, where a huge dusty disc of material surrounds a young star. This is similar to the cloud of material from which the Earth and other planets are thought to have formed around the young Sun, and it can be studied in detail because it is relatively close to us – a mere 375 light years away. The disc contains high concentrations of hydrogen cyanide and acetylene. When these two compounds, plus water, are used in laboratory experiments mimicking space conditions, they react to produce amino acids. And in 2019, NASA scientists announced that analysis of data from the Cassini space probe showed the presence of similar building blocks of life in active vents spewing water up from the ice-covered oceans of Enceladus, one of the moons of Saturn. These get their energy from hydrothermal sources deep beneath the ice. Powerful hydrothermal vents eject material from the moon's core; this mixes with water from the moon's huge ice-covered ocean before it is released into space through great geysers bursting through the ice as water vapour and ice grains. The molecules condense onto ice grains, where the detectors on Cassini showed them to be nitrogen- and oxygen-bearing compounds like those seen in the dusty discs around young stars. As Stanley Miller proved in the course of his long career, provided there is an input of energy it is not only easy but inevitable that molecules as complex as amino acids will be built up from simple compounds.

As the building blocks of proteins, amino acids are half of the story of life. The other half concerns the nucleic acids, DNA and

RNA. These have not yet been detected in space. But once again, their building blocks have.

The core component of both nucleic acids is a sugar called ribose. Ribose molecules are each built around a ring of five atoms, four carbon atoms and one oxygen atom, which can link up with other things outside the ring. In ribose, three of the carbon atoms are each linked to one hydrogen atom and one OH group outside the ring. But in deoxyribose, one of the three carbon atoms is attached just to two hydrogen atoms, so that the molecule has one less oxygen atom overall. Deoxyribose is ribose with one less oxygen atom, hence the name.*

Sagittarius B2 contains the building blocks of the nucleic acids among its storehouse of chemical compounds. Molecules of the sugar glycoaldehyde ($HOCH_2$-CHO) are among the compounds found in the cloud, and these are known to react eagerly with other carbon compounds to form ribose. It is perhaps a slight exaggeration to say that we have found the building blocks of RNA and DNA in space, but we have certainly found the building blocks of the building blocks – and in a significant development announced in 2019, a team headed by Yasuhiro Oba announced that they had manufactured the components of DNA in a laboratory experiment designed to mimic the conditions that exist in interstellar clouds. As Jim Lovelock, originator of Gaia theory, has put it: 'It seems almost as if our Galaxy were a giant warehouse containing the spare parts needed for life.' But even if the spare parts needed for life exist in

---

* I go into more detail about the structure of DNA and RNA in my discussion of Pillar Six.

profusion in space, and in particular in rings of dust around stars like IRS 46, how could these building blocks have got down to Earth when our planet was young?

The ice that covers grains of dust in space and provides a place where organic molecules can grow is also a key to how planets like the Earth can form. When a star forms by collapsing from a huge cloud of gas and dust as gravity tugs the material together, some dust gets left behind in a ring, like the one around IRS 46. Such a collapse is never perfectly symmetrical, because everything is rotating one way or another, so the dust settles down into a ring orbiting the parent star. If it were only dust, it would probably stay like that. But because they are covered in ice, the grains are tacky and tend to stick together when they collide, building up larger and larger lumps until they become big enough for their own gravity to pull other grains onto them. Then, the lumps can get together to make chunks of rock, colliding and merging, building bigger chunks and growing to become planets. The final stages of this process are extremely violent, with planetoids perhaps as big as Mars smashing into one another to make full-blown planets in the form of balls of molten rock. By that time, all the organics in the original grains that formed the planet – let's call it Earth – have been destroyed by the heat. But even after the Earth had formed there were still huge lumps of rocky material, many containing large amounts of ice of one kind or another, and dusty stuff around the young Sun.

The icy lumps became comets, and both comets and rocky lumps containing little or no ice were flung by Jupiter's gravity into elliptical orbits which took them through the inner part of the Solar

System where the young Earth, sterile and lacking an atmosphere, was solidifying and cooling. The result was a huge number of impacts on the surface of the Earth, so extreme that astronomers refer to it as the Late Heavy Bombardment, or LHB. Among other things, the LHB was responsible for the battered appearance of the surface of the Moon, which was already orbiting the Earth at that time. Analysis of the pattern of lunar cratering and dating of Moon rocks reveals information about the LHB, which lasted for a few hundred million years, until most of the debris in the inner Solar System left over from the formation of the planets was used up. It ended a bit less than 4 billion years ago. In less than another 200 million years, protein- and nucleic acid-based life was established on Earth, thanks to a more gentle rain of material from space that continued to fall on our planet in the aftermath of the LHB.

Comets brought both water and life – or at least the precursors of life – down to Earth. Computer simulations of these events tell us that about ten times as much water as there is in the oceans today, and a thousand times as much gas as there is in the atmosphere today, would have been released during the cometary bombardment. This helped to cool the planet, while some of the volatile material, things like water, carbon dioxide and methane, escaped into space. But as the surface of the Earth was churned up by the impacts – a process graphically referred to as 'impact ploughing' – some of the material combined with the original material of the surface to form the rocks rich in volatiles which are typical of the Earth's crust today. Once there was an atmosphere and oceans, the Earth was ready for life. And it was immediately seeded with the ingredients of life.

As well as the comets that smashed violently into the young Earth, there were many more similar objects passing through the inner part of the Solar System and gradually being evaporated away by the heat of the Sun. This is the process which gives comets their characteristic tails today, although there were many more comets, with even more impressive tails, when the Earth and Solar System were young. After 4 billion years, most of the inner Solar System comets have long since boiled away. But that is why we are here. The tail of a comet is a stream of gas and dust escaping from the icy nucleus as the comet evaporates. That dust is left in a trail around the orbit of the comet, and even today the Earth often passes through such a stream of cometary dust, producing showers of meteors as tiny particles, about the size of grains of sand, burn up in the atmosphere. But there are also particles with a more open structure, like snowflakes, which settle down through the atmosphere of the Earth and reach the ground. They carry with them the same mixture of organic material that is seen (by spectroscopy) in comet tails, and which laces the giant clouds from which planetary systems form. Samples of this material have been collected using high-flying aircraft and stratospheric balloons. The amount collected shows that even today this process is delivering about 300 tonnes of organic matter – polyatomic molecules containing carbon – to the surface of the Earth each year.

As Darwin pointed out, there is no chance for this material to develop into life today. For a start, much of it is destroyed by reactions with oxygen in the atmosphere, and the rest will get into the food chain of living things. But there was no oxygen, and no living things, when the Earth had just cooled and gained an ocean and atmosphere. So how much of this cometary manna was there to kick-start the emergence of life?

Astronomers get a rough idea from things like studies of craters on the Moon, analyses of the orbits of comets today, and computer simulations of the dynamics of the young Solar System. According to their estimates, in a span of about 300,000 years starting from the end of the Late Heavy Bombardment, as much organic material fell to Earth as there is in everything alive on Earth today. In the time from the end of the LHB to when we know for sure there was life on Earth, about 200 million years, if all the organic matter that fell could have been preserved and spread out evenly over the surface of the planet, it would have made a layer containing 20 grams of organic stuff – up to and possibly including amino acids and ribose – on every square centimetre of the surface. This is equivalent to the contents of a 250g tub of butter on every 3.5 × 3.5 cm square of the Earth's surface. No wonder life got started so quickly – and once the first life got going, it would have had plenty of stuff to feed on in the early millennia.

This is now a pillar of science. The young Earth was seeded with the raw materials of life from the cosmic warehouse referred to by Lovelock. But there is a more speculative idea, which is today maybe somewhere between stages (i) and (ii) of Haldane's classification. There is no reason to doubt that amino acids and (probably) ribose exist in interstellar clouds. Could things have gone a stage further, to produce proteins and nucleic acids inside comets? The idea is not as crazy as you might think, because the trigger which collapses a cloud of gas and dust to form stars and planets is often a supernova, the explosion of a star. This produces radioactive elements, and in the cloud icy lumps of material laced with radioactive elements could get warm enough to melt water in their hearts through the heat generated by radioactive decay. Could Darwin's warm little ponds have existed in these lumps of ice even before the Earth formed? You can decide for yourself whether this speculation is (i) worthless nonsense; or (ii) an interesting, but perverse, point of view. But if true, it implies that at least in our immediate cosmic neighbourhood, life on other planets will be based on the same kind of proteins and nucleic acids that we are based on.

Even without going that far, however, we can be sure that any Earth-like planets out there will be seeded with similar precursors of life to those that we know exist in interstellar clouds. It is hard to see how life could fail to get started under those conditions – we don't know exactly how the step from non-living to living occurs, but the fact that it occurred so swiftly on Earth suggests that it is not difficult. This is powerful evidence that Giordano Bruno was right – there really may be a profusion of planets like our own, each one inhabited by life forms built up from the same material that we are made of.

Which raises another question. How did the atoms that make up organic molecules – things like carbon, nitrogen, oxygen and hydrogen – get into clouds of gas and dust in space? Those supernova explosions provide part of the answer. But before they could play their part, complex nuclear reactions had to take place inside stars, and those reactions hinge upon another of the pillars of science, a coincidence so astonishing that it almost beggars belief.

PILLAR

The Carbon Coincidence

A combination of spectroscopy and an understanding of the physics of stellar interiors tells us that a star like the Sun is composed almost entirely of hydrogen and helium, with just a smattering of heavier elements (Pillar Two). In the heart of a star, these elements are not in the form of gases, as they would be on Earth today. The electrons have been stripped from their nuclei, which are squeezed together at enormous densities, without the empty space that makes up ordinary atomic matter (Pillar One). Observations of clouds of gas in space tell us that they have a similar composition, although there the elements are in their familiar atomic state, with the dust that is so important to life as we know it a barely significant fraction of the total amount of stuff in a galaxy like our Milky Way. There may be other things that contribute to the overall mass of the Universe, things called Dark Matter and Dark Energy, but they are outside the scope of the present book. What matters here is the kind of stuff we are made of, the chemical elements we learned about in school, referred to by physicists as baryonic matter. Where does it come from?

There is a wealth of evidence that the Universe as we know it emerged from a very hot, very dense state, known as the Big Bang, about 13.8 billion years ago. The evidence comes partly from observations that the Universe is expanding today, so that it must have been more compact in the past, partly from studies of radio noise left over from the primordial fireball (the so-called Cosmic Microwave Background Radiation) and partly from our understanding of the laws of physics. Basic physics tells us that the first baryonic matter produced from the energy of the Big Bang, in line with Einstein's famous equation, would have been hydrogen, the simplest and lightest element. The equations also tell us that as the Universe expanded and cooled about 25 per cent of that hydrogen would have been converted into helium by nuclear fusion reactions while the young Universe was still hot. But after about three minutes the fireball in which the Universe was born would have cooled to the point where no more nuclear reactions could take place, leaving great clouds made of a mixture of hydrogen and helium, the raw material of the first stars and galaxies, moving apart from one another in the expanding Universe. It doesn't take a great intellectual leap to realise that the other elements must have been manufactured later on, inside stars. But how?

To put things in perspective, and see just how much (or how little!) stuff we are talking about, we can look at the composition of the Solar System, which is representative of what we might expect to find in planetary systems orbiting other stars. As we saw earlier, in terms of mass the Sun is 71 per cent hydrogen, 27 per cent helium, and less than 2 per cent of everything else put together. In terms of the number of atoms, hydrogen makes up 91.2 per cent of the Sun, helium 8.7 per cent, and everything else just 0.1 per cent. But when the Sun was young, a lot of the lighter stuff was blown away from the dusty disc in which planets formed by the heat of the young star. The planets, and ourselves, were made from what was left over. Taking the Solar System as a whole, because some of the light stuff has been lost, in terms of mass, hydrogen contributes 70.13 per cent, helium 27.87 per cent, and oxygen, the third most common element by mass, 0.91 per cent. Although hydrogen is important in the chemistry of life (remember CHON), there is no mystery about its origin, so we can put it to one side and look at the composition of the 2 per cent of the Solar System made of relatively heavy elements, where the quantities are so small that it makes sense to talk in terms of numbers of atoms rather than mass.

Taking just the top ten elements, but not attempting to give exact numbers for the quantities of hydrogen and helium (the top two), for every 70 atoms of oxygen there are 40 atoms of carbon, nine atoms of nitrogen, five atoms of silicon, four atoms each of magnesium and neon, three atoms of iron, and two atoms of sulphur. There are only five other elements (aluminium, argon, calcium, nickel-iron, and sodium) which have abundances between 10 per cent and 50 per cent of the abundance of sulphur. All the even heavier elements are very much rarer. For every 10 million atoms of sulphur there are, for example, only three atoms of gold, which is one of the reasons that gold is valuable, and which tells us something profound about the Universe which I shall come to shortly.

The first clue to how the elements are made inside stars comes from that list of the top ten – or at least, the members of the top ten heavier than helium. The nucleus of a helium atom (strictly speaking an atom of helium-4) is identical to an alpha particle, made up of two protons and two neutrons. A carbon nucleus is made up of six protons and six neutrons, like three alpha particles stuck together, which give it the name carbon-12. Adding another alpha particle gives you oxygen. Nitrogen, silicon, magnesium, neon and iron all have nuclei which contain whole numbers of alpha particles. If alpha particles can be added to nuclei inside stars, they will build up exactly this chain of elements. Rarer elements can be produced by occasional nuclear processes involving stray particles such as electrons, neutrons and protons interacting with the most common nuclei. This build-up of heavier elements can happen because the balance of energy involved favours heavier (more massive) nuclei over lighter ones, as with the conversion of hydrogen into helium, all the way up to iron. A nucleus of carbon-12, for example, is slightly less massive than three alpha particles, and if three alpha particles are combined (by whatever means) into one carbon-12 nucleus, the 'lost' mass is released as energy. Similarly, in terms of the overall energy, an oxygen nucleus is a more efficient arrangement than a carbon nucleus with a separate alpha particle, and so on up to iron. Even heavier elements are a separate puzzle, because their nuclei are less efficient packages of mass-energy, so it requires an input of energy to force nuclei to squeeze into one another to make elements such as gold. But first things first. In the 1940s, when the pioneering astrophysicist Fred Hoyle tackled the problem of what became known as stellar nucleosynthesis, he started with the puzzle of making everything up to iron inside stars.

Nuclear fusion releases energy when lighter nuclei join together to make heavier nuclei, up to iron. But all nuclei have a positive electric charge, and are repelled from one another by electrical forces. They can only fuse if they are squeezed so tightly together that nuclear forces overwhelm the electric force which tries to keep the nuclei apart. This means they have to be moving very fast when they collide, and their speed is related to the temperature. By the mid-1940s, physicists had a good idea of the temperatures required to do this for different fusion reactions, but there was a big problem with one of the first steps in the process of building up nuclei by adding alpha particles.

You may have noticed I haven't mentioned any nucleus composed of two alpha particles. The element this corresponds to is called beryllium-8, but it is never found in nature. Beryllium-8 nuclei are unstable, and if they are manufactured artificially they fall apart almost instantly. A couple of astrophysicists suggested that the way to get across the gap between helium-4 and carbon-12 would be if three alpha particles came together simultaneously inside a star, fusing to make a single nucleus of carbon-12 without beryllium-8 having been formed along the way. But such a triple collision would involve so much kinetic energy that it would be more like a train wreck than a smooth fusing of alpha particles. How could such a process proceed smoothly?

Hoyle's insight started with the realisation that there is no need for the three alpha particles to collide literally simultaneously. Although the lifetime of beryllium-8 is small – each nucleus lasts

Fred Hoyle

A. Barrington Brown © Gonville & Caius College/Science Photo Library

for only about 10^{-19} seconds – under the conditions that exist at the heart of a star there are so many alpha particles that collisions are constantly manufacturing them. There are always some around, just as there is always some water in a sink being fed by an open tap while water is escaping from the plughole. In a star with a central temperature of around 100 million degrees C, about one nucleus in every 10 billion will be beryllium-8. So there is always a population of beryllium nuclei which are 'targets' for alpha particles, opportunities to make carbon-12 nuclei. But even this prospect didn't look promising, because it couldn't make enough carbon to explain the amount seen in the Universe unless another factor was at work.

In 1953, Hoyle realised what the other factor was. All nuclei can exist in different energy states, called resonances. The usual analogy is with a plucked guitar string. This has a fundamental note, but it can also play different harmonics of that note. Nuclei have a basic energy level (the ground state), but if they are given extra energy they can jump up to an 'excited' state, like a ball being lifted up a staircase to a higher step. Like the ball bouncing down the stairs, such excited nuclei will soon give up the extra energy (perhaps in the form of gamma rays) and settle back into the ground state.

Hoyle calculated that under the conditions inside a star, other things being equal the collision of an alpha particle with a short-lived beryllium nucleus would simply blow it apart. But he reasoned that if the energy of the incoming particle was just right, it would nudge the combined nucleus gently into an excited state of carbon-12, like a ball being gently placed on a high step of a staircase, from which it could radiate energy and jump down into the ground state of carbon-12. The snag was, the trick only worked if there was an excited state of the carbon-12 nucleus at a very precise energy level, 7.65 million electron Volts above the ground state, in the units used by physicists. If the energy level was even 5 per cent higher the trick would not work. And nobody knew if there was such an excited state of carbon-12 at all.

Nobody took Hoyle's idea seriously. But his argument seemed to him to be watertight. Carbon exists in the Universe. Indeed, we are made partly of carbon. It has to have been made somewhere – and where else could it have been made other than inside stars? At the time Hoyle, although based in the UK at Cambridge, was visiting the California Institute of Technology, and took the opportunity to ask an experimental physicist, William Fowler, to carry out an experiment to test his idea, looking for the predicted resonance of carbon-12. Actually, he did more than request. He badgered Fowler into submission. Fowler told me that he thought Hoyle was crazy, but he eventually agreed to put a small team together to do the experiment to shut him up. Whatever the motivation, the experiment got done. It took three months, and it proved Hoyle was right. There is indeed a carbon resonance in just the right place to explain how the 'triple alpha' process works. Everybody was surprised – everybody except Hoyle.

This is one of the greatest triumphs of science, perhaps the single most significant example of a theory making a prediction and being proved right by a laboratory experiment. It was well worthy of a Nobel Prize, but Hoyle never received one, although Fowler did, for the work which both of them and two other colleagues developed from this beginning.

On his next visit to Caltech, Hoyle got to know the British husband-and-wife team Geoffrey and Margaret Burbidge, who were temporarily based in California (they eventually moved there permanently) and were trying to understand the significance of the exact abundances of various elements in stars, as revealed by spectroscopy. Fowler got roped in to this work as well, with the team calculating how a steady supply of neutrons inside stars could convert nuclei produced by the alpha process into other elements in the proportions actually observed. Hoyle initially kept in touch with the work from a distance, but in 1956 all four were together in California, where they put everything in place in a massive scientific paper that was published in the October 1957 issue of the *Reviews of Modern Physics*. The names of the authors of this masterwork are listed alphabetically, as Burbidge, Burbidge, Fowler and Hoyle, and to this day it is referred to simply as B^2FH.* But everybody knew that the guiding inspiration for the work came from Hoyle – everybody except the Nobel Foundation, which eventually, in 1983, gave Fowler alone the prize for this breakthrough. Fowler was embarrassed, but accepted the award. When Fowler died, Geoffrey Burbidge referred to the decision in his obituary of his old friend, saying that this award 'caused some strain among B^2FH, since we were all aware that it was a team effort and the original work was done by Fred Hoyle'. But the work stands as a pillar of science, whoever got the recognition. Without this coincidence between the carbon resonance and the amount of energy carried by a fast-moving alpha particle inside a star, there would be no carbon,

---

* Pronounced 'B-squared F H'.

no heavier elements, no complex molecules in the gas clouds from which stars form, no planets like Earth, and no life forms like us in the Universe.

This work essentially explained how all of the elements are manufactured inside stars, up to iron-56 and nickel-56 (iron-56 contains 26 protons and 30 neutrons in each nucleus; nickel-56 has 28 protons and 28 neutrons in each nucleus, fourteen alpha particles fused together). The manufacture of even heavier elements involves some of the most violent events seen in the Universe today, when whole stars explode as supernovae. Fowler and Hoyle (speaking alphabetically) were also involved in developing this understanding of stellar nucleosynthesis. But that understanding has since been extended to take on board studies of even more violent events.

The supernova connection involves stars much more massive than our Sun. For stars with masses about one to four times the mass of our Sun, after the conversion of hydrogen to helium in its core the star shrinks a little, gets hotter in the middle, and 'burns' helium into a mixture of carbon and oxygen. But that is as far as it goes. During the later stages of its life, the star blows a lot of material, including carbon and oxygen, out into space, then settles down as a white dwarf – a cooling cinder with a mass close to that of the Sun today, but no bigger than the Earth. More massive stars lead more interesting, and spectacular, lives. The extra mass is important because more inward pressure is required to get the interior of the star hot enough for successive phases of nuclear burning to take place. Carbon is converted into neon, sodium and magnesium, by the processes studied by B^{2}FH, at a temperature of about 400 million degrees C; oxygen burning produces silicon, sulphur and other elements at a temperature of about 1,000 million degrees C. The silicon-28 (effectively seven alpha particles stuck together) produced in this way is eventually converted into iron and nickel. But at each stage of the process, a residue is left behind, so that a massive star at the end of its life contains a core of hydrogen, surrounded by a shell of helium, surrounded by successive shells of other elements nested like onion skins.

When all its sources of nuclear energy are exhausted, the star will collapse. But this releases gravitational energy, generating so much heat that the star explodes as a supernova. Some of the explosion goes inward, compressing the core of the star and turning it into a neutron star (with as much mass as our Sun squeezed into a ball about 20 km across), or even a black hole. But a great deal of the blast goes outwards. It provides the energy which manufactures elements heavier than iron in the outer part of the star, and also spreads these elements, and the others built up during the life of the star, out across space to form the raw material for new stars and planets – and, on at least one of those planets, people.

All this was clear by the end of the 1960s, although many details were filled in over the decades that followed. But there was a nagging problem. Even though the traces of very heavy elements such as gold seen in the Universe are small, the ever-improving calculations and computer simulations showed that a supernova explosion could not make enough of them to explain the observations. Matching up the observed rate of supernova explosions with the observed amount of things like gold, platinum and uranium in the Universe, scientists found that only half of the very heavy elements could be accounted for in this way. Something else was required to make the rest, and without knowing exactly what it might be, astronomers gave it a name – a kilonova. Completing the story of the origin of the elements, and confirming the accuracy of those calculations dating all the way back to Hoyle's insight, kilonova explosions were detected at last in 2017, but not (initially) by their light.

On 14 September 2015, astronomers opened a new window on the Universe. For the first time, they detected gravitational waves – ripples in space – from a violent event far away across the Universe. That event was the merger of two black holes. The discovery of gravitational waves had long been anticipated – they are a prediction of Einstein's general theory of relativity – and long sought. But they are incredibly tiny, by the time they reach Earth, and very hard to observe. The 'telescopes' that made the observation were based around evacuated tubes 4 km long, in which mirrors that reflected laser light to and fro along the tunnels were so delicately balanced and so precisely monitored that when the mirrors moved across a distance less than the diameter of an atom, the wobble could be measured.* Einstein's theory predicts exactly what kind of wobble would be produced by waves from things like merging black holes, and that kind of wobble was precisely what was detected in September 2015. Since then, the detectors around the world (there are now two in the United States and one each in Europe and India) have found several other gravitational wave 'events', as the astronomers like to call them, and one in particular is relevant to my story.

....

* For details, see https://www.amazon.co.uk/Discovering-Gravitational-Waves-Kindle-Single-ebook/dp/B071FFJT74

On 17 August 2017 the detectors picked up a slightly different pattern of ripples, lasting for just 100 seconds, which matched the predictions of the pattern that would be produced when two neutron stars collided with one another. This was especially exciting because unlike the merger of two black holes, a neutron star collision was expected to produce an explosion of light and other radiation, such as gamma rays. Neutron star mergers were, indeed, probable candidates for the hypothetical kilonova explosions in which very heavy elements might be formed, and astronomers had calculated, based on the number of stars around in a galaxy like our own, how common such events might be.* The direction the gravitational waves had come from was indicated approximately by the observations, and within hours of the detection astronomers were pointing their telescopes in that direction. They found a short-lived bright object in a nearby galaxy called NGC 4993, roughly 130 million light years away from us. It was a kilonova. Spectroscopy showed that the kilonova had indeed produced a lot of heavy elements, such as uranium, gold and platinum. This included 200 times the mass of the Earth in the form of gold, and 500 Earth-masses of platinum. When the amount seen in this explosion was multiplied by the calculated frequency of neutron star mergers, the result was that such explosions could indeed produce the 'missing' half of the heavies. Among other things, this means that if you have a wedding ring or some other item made of gold or platinum, you can be sure that very many of the atoms in this object were manufactured during the collision of

---

* Because neutron stars are so dense, these collisions are very efficient at making heavy elements, but they only produce about a tenth as much light as a supernova, so they are harder to find.

two neutron stars, and spread into space in a gigantic explosion, seeding the cloud from which the Sun and Earth formed.

So we know how the elements are made in stars, we know that those elements are combined into complex organic molecules in space, and we know that these complex molecules were brought gently down to the surface of the Earth soon after it formed, where they became the key components of life. But how do those components work together to make things like us? The answer involves another surprising pillar of science.

PILLAR

The Book of Life is Written
in Three-letter Words

The complexity of life is built from two families of molecules – proteins and nucleic acids. These molecules are themselves built from a relatively modest variety of compounds. There are 92 elements that occur naturally on Earth, but just 27 of them are essential for living things, and not all of the 27 are found in all living things.

Proteins have two roles to play. One kind provides the structure of the body – things like hair, muscle, feathers, fingernails and shells. A combination of X-ray analysis, chemistry, and an understanding of the quantum mechanical processes that hold atoms together to make molecules has shown that these proteins are made from long chains of amino acids, forming a helical structure. It's fairly obvious that this kind of molecule can produce long, thin things like hair, but it can also produce hard sheets of stuff like fingernails when the individual helices are linked together side by side by chemical bonds of one kind or another. All of this was established by Linus Pauling and his colleagues at the California Institute of Technology, who published a groundbreaking series of seven papers on the structure of proteins in the journal *Proceedings of the National Academy of Sciences* in 1951. The other kind of protein provides the workers of the body. Things like the haemoglobin that carries oxygen around in your blood, and substances known as enzymes that encourage (or in some cases inhibit) certain chemical reactions that are important to life. Unravelling their structure proved a tougher nut to crack.

A clue to why this kind of protein is hard to deconstruct comes from the name they were eventually given – globular proteins. It turned out that they are also made of long chains of amino acids, but that in this case the chains are curled up into little balls, and each kind of globular protein has its own distinctive three-dimensional shape. It is the shape of a globular protein, as much as its chemical composition, which determines its role in the chemical processes of life. For example, haemoglobin has a cavity which is just the right size and shape for a molecule of oxygen to nestle in. Or think of a globular protein which has two indentations, each just right for a different smaller molecule to sit in. When they do so, they will be aligned in such a way that bonds can form between them, before they are released as a single larger molecule. This is reminiscent of the way small molecules got together on the surfaces of icy grains in the depths of space before the Earth formed. One enzyme may, for example, mindlessly and repeatedly join together specific pairs of amino acids to make one link in a growing chain that will become another protein molecule.

The structure of haemoglobin itself was worked out by 1959, by

researchers funded by the UK Medical Research Council laboratory. It is made up of four chains, each made of similar amino acids,
locked together to make a roughly spherical ball which actually has
four pockets on its surface where oxygen molecules can nestle. And
almost identical chains are found doing exactly the same job in the
blood of creatures as different as horses and whales. Evolution is
very conservative – once it has found a molecule that is good at
doing a specific job, it sticks with that molecule without replacing it.
But how does it know how to make these molecules? This is where
nucleic acids come in to the story, although it took a long time for
the role of DNA and RNA to be appreciated.

When the nucleic acids were first identified as major components of living cells, it was thought that they were some kind of
structural material, like scaffolding, to which the much more complex and (it was thought) more interesting protein molecules were
attached. It was a simple mistake to make, because on the face of
things DNA and RNA molecules are themselves simple. Each of
them is a long molecule made up of four sub-units called bases.
Three of these are the same in both RNA and DNA; the fourth base
is different in the two molecules, so there are five bases involved in
all. They are uracil (U), thymine (T), cytosine (C), adenine (A) and
guanine (G). DNA molecules contain G, A, C, and T, while RNA
molecules contain G, A, C, and U. U, T, and C are built around six-
sided rings of carbon and nitrogen atoms, while A and G are based
on two such rings joined side by side, like a figure 8. These bases
are attached to a backbone containing the relevant sugars (ribose
or deoxyribose) linked together in a chain, with the bases sticking
out to the side of the backbone. The details of this were not known,

Raymond Gosling
King's College London Archives/Science Photo Library

however, until the early 1950s, when Francis Crick and James Watson in Cambridge used X-ray data obtained by Rosalind Franklin and Raymond Gosling at King's College in London, which had been passed to Watson by a colleague without their knowledge or permission, to determine the famous 'double helix' structure of DNA. The original idea was that the bases were laid out in a regular way along pieces of scaffolding – something like GACTGACTGACTGACT ... in DNA and GACUGACUGACUGACU ... in RNA. This is not a 'message' that conveys much information.

That was more or less where things stood in the mid-1940s. It was known that the genetic material that passes on the blueprint, or recipe, for life is contained in large structures called chromosomes found in the hearts of cells, and that these chromosomes are copied and passed on to later generations to carry the recipe forward. But chromosomes were known to contain both DNA and proteins, and the proteins were thought to be the important component for conveying information. One way in which a cell might 'know' how to make the proteins it needed to function might be, for example, if one sample of each protein was attached to the scaffolding of DNA, ready to be copied when required. It made sense, but it was wrong.* Even so, it was clear that chromosomes carry the recipe of life, in some form of 'code'.

The person who set scientists on the trail of that code of life was a physicist, Erwin Schrödinger, best known today as the originator of the famous 'cat paradox' of quantum mechanics.[†] In 1943,

----

* Almost the opposite of the truth; in chromosomes, proteins are the structural material and DNA carries the information, as I shall explain.
[†] See *Six Impossible Things*, pages 47–49.

Schrödinger was based at the Institute for Advanced Studies in Dublin, where he had moved as a refugee from the Nazis after their takeover of Austria. That year, he gave a series of lectures at Trinity College on the theme 'What is Life?' They were published the following year in a book with the same title. This book would have a huge influence on the generation of scientists who set out to crack the code of life after the end of the Second World War, including Crick and Watson.

The key idea that Schrödinger introduced and passed on to those researchers was that 'the most essential part of a living cell – the chromosome fibre – may suitably be called *an aperiodic crystal*'.* He thought that the key component of a chromosome was protein, but that doesn't matter, because his insight works equally well for fibres made of DNA. A periodic crystal, to use his terminology, would be something like the structure of common salt, sodium chloride, in which alternating atoms of sodium (Na) and chlorine (Cl) form a repetitive array in three dimensions, NaClNaClNaClNaCl ... which has a structure but conveys very little information. This is very similar to the idea of DNA as the scaffolding on which proteins might be hung. What Schrödinger meant by an aperiodic crystal can be understood in terms of a tapestry. If you had some strands of a few colours of thread, they might be arranged side by side and woven to make bands of colour – for example, red, yellow, blue and green – in a striped blanket. This would be equivalent to a periodic crystal. Or the same threads could be woven in a more complicated way to make a picture of a flower. This would be equivalent to an

----

* His emphasis.

aperiodic crystal. Schrödinger pointed out that although it is made up in this way from strands with just a few different colours, there is a structure in 'a Raphael tapestry which shows no dull repetition but an elaborate, coherent, meaningful design'.

Schrödinger also pointed out that what he referred to as a 'code-script' carried by an aperiodic crystal in the chromosome fibres could contain all the information required to make proteins, without it being necessary for a copy of each kind of protein to be carried as a template in the chromosomes themselves. Just 20 different amino acids are required to make all of the different proteins important for life, and if you think of these amino acids as 'words' strung out along a protein molecule to make a sentence (or a book!) you would have nearly as much scope for conveying information as the 26 letters in the English alphabet which I am using to write this book, which (I hope you agree) contains a lot more information than a boring repetition of the alphabet from A to Z. But would you need even a 20-letter alphabet to write the book of life?

There is no need, Schrödinger realised, for anything as complicated as amino acids. Even individual atoms could do the job if they could be organised properly: 'The number of [different] atoms in such a structure [the aperiodic crystal] need not be very large to produce an almost unlimited number of possible arrangements.' He gave as an example the Morse code, where there are just two basic signs, dot and dash, but which can be put together in groups of up to four symbols to make 30 different specifications, sufficient for the English alphabet plus a few punctuation marks. And with a third sign, using them in groups of not more than ten symbols, 'you could form 88,572 different "letters"; with five signs and groups up to 25, the number is 372,529,029,846,191,405'. Schrödinger got a bit carried away by his training as a physicist here, since there is no need for such a fantastically high number of words. But that really only became clear after the structure of DNA had been determined.

That story is too well-known for it to be necessary to go into detail here, but what matters is that each molecule of DNA is composed of two strands, twining around each other in the famous double helix. Each single strand of DNA has a spine made of a chain of sugar groups linked together by phosphate groups (a phosphate group is made of a phosphate atom surrounded by four oxygen atoms). As we saw earlier, the bases (G, A, T, and C) are attached to the sugar groups, and stick out from the sides of the spine. Different pairs of these bases have an affinity for one another, thanks to their shape and to a weak form of electric attraction called the hydrogen bond. Thymine and adenine naturally link together in this way, as do cytosine and guanine. This holds the two strands of DNA together, but relatively loosely. Everywhere on one strand that there is T, on the opposite strand there is A; everywhere on one strand there is C, on the opposite strand there is G. And vice versa. This pairing was the key to the Crick–Watson model of DNA, and at the end of their famous paper, published in *Nature* in 1953, they rather coyly stated:

It has not escaped our notice that the specific pairing we have postulated immediately suggests a possible copying mechanism for the genetic material.

Which was their ploy to establish their priority* for the idea that DNA can be replicated if the two strands untwist and then each strand builds a new partner for itself by hooking up with other components from the chemical soup inside the cell. Every A on a single strand captures a T from the soup, every T captures an A, every G captures a C, and every C captures a G. The result is two identical double helices where there used to be one. Genetic material has been copied. Exactly how the mechanisms of the cell do this was far from being understood in 1953, but the important point was that clearly this was something that could work, in principle. The big question this raised was, what was it that was being copied? How did DNA store the information in the book of life?

It was another physicist, the Russian-born American George Gamow, who set people, in particular Francis Crick, on the trail. He later recalled that in 1953, while a visitor at the Berkeley campus of the University of California:

> I was walking through the corridor in Radiation Lab, and there was Luis Alvarez going with *Nature* in his hand ... he said 'Look, what a wonderful article Watson and Crick have written.' This was the first time that I saw it. And then I returned to Washington and started thinking about it.[†]

Gamow came up with the idea that protein molecules could be built up directly along strands of DNA, if the row of bases along the DNA

---

* The ploy worked, as the fact that I am quoting it here proves.
[†] Interview in the George Gamow Collection of the Library of Congress, Washington, DC.

carried the code for each amino acid required for the protein in the right order along the DNA molecule. This echoed Schrödinger's idea, which he was unaware of. Gamow wrote to Watson and Crick with his idea, and spelled it out in a paper in *Nature* published in 1954:

> The hereditary properties of any given organism could be characterised by a long number written in a four-digital system. On the other hand, the [proteins] are long peptide chains formed by about twenty different kinds of amino acids ... the question arises about the way in which four-digit numbers can be translated into [amino acids].

The details of Gamow's idea were wrong, but by talking about the code of life in this way he prompted Crick and many others to try to work out how such 'translation' might work. A key step was understanding the role of the other nucleic acid, RNA.

One puzzle about how DNA could be actively involved in the workings of a cell is that the DNA is packed away at the heart of the cell, in its nucleus. All the action, including manufacturing protein, takes place in the outer part of the cell, the cytoplasm. There is very little DNA out there but plenty of RNA. And although the amount of DNA in every cell of a particular organism is the same for every cell and all the time, the amount of RNA varies considerably from cell to cell and from time to time in any individual cell. It became clear that it is RNA that is directly involved in making protein, and that bits of genetic code are copied from the DNA onto new strands of RNA as required, then the RNA is released out into the cytoplasm and used to build up protein molecules roughly in the way Gamow suggested, after which the RNA strands are broken up and the parts reused. DNA is like a library, a storehouse of information, from which individual books, instruction manuals for the manufacture of specific proteins, are copied onto RNA as required. When part of a DNA molecule in the nucleus is uncoiled and copied onto RNA using the mechanism which had 'not escaped the notice' of Crick and Watson, each T is replaced by a U, but there is no other significant difference. So for simplicity, when discussing the genetic code from now on I shall describe it in terms of the RNA bases, U, C, G, and A.

Cracking the code involved a lot of people carrying out a lot of biochemical investigations, unravelling the details of the workings of the cell step by step. But this is not the place to go into all those details, which you can find in Horace Judson's book,* and I shall focus on the thinking behind the experiments, and the conclusions resulting from the experiments. Very early on, the researchers decided to concentrate on a triplet code, not the four-digital code suggested by Gamow, because three letters are all you need. If you have four bases and treat each of them as a letter, then using each one individually you can only code for four amino acids. With two bases at a time in doublets, you can manage sixteen different arrangements – sixteen words, not enough to code for the 20 amino acids essential for life. But with three-letter words, or triplets, you can manage 64 different combinations, more than enough to code for all the necessary amino acids, with some left over to act as the

---

* See Further Reading, page 245.

equivalent of punctuation marks, including markers for the begin-ning and end of a particular 'message'. With a four-letter code, the number of individual words would be 256, far more than required.

During the 1950s and 1960s, biochemists carried out experi-ments involving strands of RNA made up of a variety of bases, to see what kind of proteins they manufactured. A key breakthrough came with the discovery that a boring strand of RNA carrying the repeat-ing chain of bases UUUUU ... (poly-U), when placed in a suitable chemical environment mimicking the inside of a cell, would make a boring chain made of repeating units of the amino acid phenyla-lanine, phe.phe.phe ... (poly-phe). This is technically a protein, but of no use to living things. It meant, however, that the first triplet word had been identified. The code UUU in RNA corresponds to the amino acid phe. A huge amount of work along these lines led to a complete understanding of the code. Every one of the triplets that can be formed out of the bases U, C, G, and A was linked with a specific amino acid or with a punctuation mark. Some of the amino acids are coded for by several triplets – for example, valine can be indicated by GUU, GUC, GUA or GUG – but this redundancy does not affect the way the book of life is read. Surprising though it may seem, the whole story of life really is written in three-letter words. But you need a lot of words to tell that story.

How big is the book? In human cells, the DNA packaged into chromosomes in the nucleus is coiled up in coils which are them-selves coiled into supercoils. There are roughly 3 billion pairs of bases, linked across DNA strands, in each of the cells, packed so tightly that they take up a space only about six microns (six mil-lionths of a metre) across. If all this DNA could be untwisted and stretched out, it would be about two metres long. And if all the DNA in all the cells in your body were stretched out in this way and laid end to end, it would stretch along about 16 billion kilometres – more than a thousand times the distance of the Earth from the Sun.

Exactly how bits of DNA are untwisted from this compact state and copied onto RNA when required is still not fully understood. But there is one key feature of the process which you may have already latched on to. It is only possible because the twin strands of a DNA molecule are only held together loosely, by the hydrogen bonds that I mentioned so casually earlier, so that they can be opened up and closed again like the opposite sides of a zip. Hydrogen bonding is a key to the existence of life as we know it, and it can be understood more easily in the context of another scientific surprise – the incredible lightness of ice.

The Incredible Lightness of Ice

I ce floats on water. This is so obvious that most of us never think about it. But it is a key feature of our environment, and it is distinctly odd, as a little home experiment indicates. If you take two see-through containers, and partly fill one with water and the other with olive oil, then put them into a freezer, the liquids will both solidify. The water makes ice, while the olive oil turns into a solid similar to butter. Now take the containers out and stand them on a warm table while they thaw. In one container, liquid water forms at the bottom as the ice melts, until all that is left is a layer of ice floating on top of the water. In the other, the lump at the bottom of the container stays solid, while molten oil rises to the surface, forming a liquid layer above the solid lump. The second situation is more representative of how things behave. Most solids are heavier (denser) than their liquid form so they sink. Why should ice be different, and how has that affected the evolution of life on Earth?

Although people had previously noticed what was sometimes called 'the expansion by cold' of water near its freezing point, the first person to carry out a proper scientific study of the phenomenon was Benjamin Thompson, Count Rumford, in the first decade of the nineteenth century. Rumford was a colourful character worthy of a book in his own right, who started life as plain Ben Thompson in the American colonies in 1753, fought on the British side in the American War of Independence, made his way after the war to Bavaria (where his many services to the Duke led to him being awarded the title of Count), made pioneering studies of the nature of heat, and founded the Royal Institution in London. Along the way, he investigated what happens to water as it approaches its freezing point.

It is typical of Rumford, who never switched off from work, that some of those investigations were triggered by observations he made while on holiday in the Swiss Alps with the beautiful Madame Marie Lavoisier (the widow of the pioneering chemist Antoine Lavoisier), whom he later married. On the surface of the great mass of ice on the Chamonix glacier, Rumford saw 'a pit perfectly cylindrical, about seven inches in diameter, and more than four feet deep, quite full of water'. After the guides told him that such pits are quite common, he reasoned out how they formed. Warm summer winds blowing over the ice could melt surface ice in gentle natural depressions. The water at the top of these puddles is a tiny bit warmer than the water lower down, so it is denser, sinks, and gives up its heat to the ice at the bottom of the puddle and melts it. The now slightly cooler water is lighter and rises to the surface, being replaced by slightly warmer water falling down, in a perfect example of inverted convection, 'by which the depth of the pit is continually increased' until the cold weather returns.

Count Rumford
Collection Abecasis/Science Photo Library

Rumford wrote all this up in a paper published in the *Philosophical Transactions* of the Royal Society in 1804, where he stressed that these studies:

> ought not to be regarded as suitable for determining with great exactness the temperature at which the density of water is at a maximum, but rather as proving that this temperature is really several degrees of the thermometric scale above that of melting ice.

But just a year later he presented a paper to the National Institute of France describing a neat experiment which did establish reasonably accurately the exact temperature at which water has its maximum density.

Rumford filled a container with ice on the point of melting, exactly at the freezing point of water. Inside the ice bath there was a second container, and inside that was a little cup-shaped receptacle, in contact with a thermometer. Directly above the cup there was a heated ball which could be dipped into the slush at the top of the ice bath, and warmed the water there. As Rumford anticipated, the warm water was denser than the icy water and flowed down into the cup; the densest water filled the cup, where its temperature could be measured. He found that the cup filled with water at 41 degrees Fahrenheit, equivalent to about 5 degrees Celsius (modern measurements give the temperature at which water has its maximum density as 4°C, so he did remarkably well with the equipment he had available).

The question this raised was, why did water behave in this way? The answer lies in the nature of the hydrogen bond, which was only properly understood after the development of quantum theory in the 1920s, but which you can get a rough idea of in general terms. It depends on the fact that the hydrogen atom is the simplest of all the elements, and has just a single negatively charged electron in some sense orbiting around a single positively charged proton.* Atoms can combine to form molecules when they share electrons with each other to form a link, and some configurations are particularly favoured by the quantum rules. For example, a hydrogen atom would 'like' to have two electrons, so it will eagerly join together with any other atom that has an electron available for pairing to make a molecule, with the two electrons shared between them. This is only possible for certain partners, because of the way the quantum rules affect the pairing. Carbon atoms, for example, can form four bonds, oxygen atoms two, and nitrogen three. But when hydrogen does form a bond in this way, with the electrons in a sense forming a bridge between a proton and another atom, the other side of each proton is exposed, with no screening of negative electric charge outside it. This means that it can form weaker links with atoms that have a surplus of negative charge, apart from the electrons used in ordinary bonding, available for this weaker form of pairing. This only happens for hydrogen atoms; nuclei of other atoms are screened by additional electrons not included in regular chemical bonding. But it is that very surplus of electrons that provides an opportunity for them to form the other end of hydrogen bonds.

Although hydrogen bonds can form between other molecules

....

* Only in some sense because quantum physics tells us that electrons do not behave purely as tiny particles, but have wave-like properties as well.

(not least in DNA, and in the links that give protein molecules their interesting and important shapes), the ones involving water are particularly strong and particularly important for us. Water molecules are made up of two hydrogen atoms and one oxygen atom, H_2O. Each oxygen nucleus contains eight protons, so there are eight electrons in the cloud surrounding the nucleus. Just two of these are involved in the bonds with the hydrogen atoms, so there are six unattached electrons in the cloud. These provide an electrical attraction for the partially exposed hydrogen nuclei belonging to nearby water molecules.*

Each oxygen atom in a water molecule can form two hydrogen bonds in this way, while on the other side of the water molecule each hydrogen atom can form a single hydrogen bond with an oxygen atom in another molecule. This makes four bonding possibilities in all, encouraging the formation of hydrogen bonds arranged in a tetrahedron around each water molecule, which produces the open crystalline structure in the solid (think snowflakes), and also encourages water molecules to tug on one another as they move about in a liquid. Which is why water is liquid at all the temperatures we find comfortable on Earth today.

Whether a substance is in the form of a solid, liquid or gas depends on the temperature, other things (in particular pressure) being equal. The higher the temperature, the more energy the particles making up the substance (the atoms or molecules) have, so the faster they move. At high enough temperatures, they fly around

* They cannot form 'proper' bonds for quantum-mechanical reasons beyond the scope of this book.

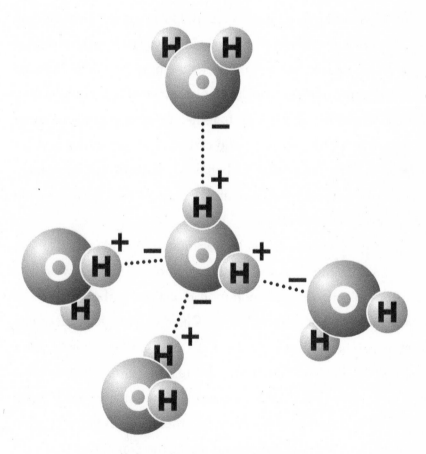

Hydrogen bonding of water molecules

freely, bouncing off each other and the walls of any container they are confined in. At a range of lower temperatures, they are almost touching, but still have enough energy to slide past one another. At still lower temperatures, they are scarcely able to move at all, except to do a kind of jogging on the spot, and form a solid. Heavier molecules need more energy to make them move faster, so by and large substances made of heavier molecules ought to melt and evaporate at higher temperatures than substances made of lighter molecules, except where the atoms link together to form crystals or other arrays, as with, for example, solid carbon. The peculiarity of water can be seen by comparing its behaviour with that of substances made up of molecules with roughly the same, or even greater, weight than molecules of water.

On a scale where a single hydrogen atom has one unit of mass, oxygen has 16 units, so a single oxygen molecule (H_2O) weighs in at 18 units. Another very common molecule, carbon dioxide, is made up of two oxygen molecules joined to a single atom of carbon, which has a mass of 12. So it has an overall mass of 44 units. Yet carbon dioxide is a gas at room temperature, while water is a liquid. Hydrogen sulphide (mass 34), methane (mass 16) and nitrogen dioxide (mass 46), among others, are all gases at room temperature. Water is only liquid under the conditions that exist at the surface of the Earth because hydrogen bonds make water molecules sticky. Even when the molecules are moving about in the form of a gas and the attraction between oxygen molecules is not strong enough to slow them down and form permanent hydrogen bonds, the hydrogen bonding effect still has an influence. In liquid water, although the distances between neighbouring molecules is large enough and the energy of the molecules is great enough that hydrogen bonds that try to form are stretched and broken, they still form temporarily. In a gas, the effect gets stronger when the temperature gets down towards the boiling point, 100°C; in liquid water itself, the molecules are closer together than they would be without the hydrogen bonding effect. And when the temperature gets close to freezing, the effect is dramatic.

Down to about 4°C things proceed roughly as you might expect, with the density increasing as the water cools and molecules move more slowly. Water at 4°C is about 4 per cent more dense than water near the boiling point. But below 4°C, the molecules are moving so slowly that they begin to arrange themselves in the tetrahedral pattern typical of ice. Even before they can form permanent crystals, this reduces the density of the liquid, just as Rumford observed. And when the solid ice does form, it floats on water. There are other substances that form spacious crystal lattices and expand on freezing, including acetic acid, silicon, gallium, and (if you want to risk it) plutonium. But water is of key importance to life on Earth, and the incredible lightness of ice may be a major reason why we are here.

There are several benefits of hydrogen bonding that may not be immediately obvious. For example, it allows animals such as us to cool down by sweating because a large amount of heat is needed to break hydrogen bonds between water molecules and make the water evaporate (it is the energy used in evaporating sweat that helps to cool us down on a hot day); and the proximity of a large body of water that can absorb heat when temperatures are high and let it out when temperatures are low reduces the range of temperature variations near the sea, keeping summers relatively cool and winters relatively warm. This is more than just a convenience for coastal dwellers today. It is the hydrogen bonding effect which allows the existence of large bodies of water even when the temperature drops below freezing, because a layer of ice on top of the water acts as an insulating blanket which keeps the water underneath the ice liquid. Without this effect, our planet might be a frozen, lifeless iceball, judging from the geological record.

If the hydrogen bond did not exist, there would, of course, be no liquid water on Earth at all. But imagine for a moment a planet cool enough for liquid water to exist without the benefit of hydrogen bonding. What would happen if it got cold enough for ice to form? The ice, being denser than liquid water, would settle to the bottom of the ocean. That would leave the top of the ocean exposed to the cold, so more water would freeze and sink to the bottom. Soon, the entire ocean, or lake, would be frozen solid. This would happen to all the water on the planet. It would be very difficult to thaw such a frozen planet, because the shiny white surface of the ice would reflect away the incoming solar heat. Life like us could not exist in such circumstances. And even with the benefit of hydrogen bonding, the Earth has been through more than one 'snowball' event during its long history.

Geological evidence in the form of scars in the rocks and the kinds of sediments deposited in the oceans at different times tells us that the Earth froze over entirely about 2.5 billion years ago, and froze again some time between 700 and 600 million years ago. There may have been other similar events, but the evidence for them is not conclusive. Nobody knows what causes such events. Speculations include vast volcanic outbursts on Earth which throw material high into the atmosphere and shield the surface from the Sun's heat, or collisions between asteroids in space that spread dust through the inner part of the Solar System to make a sunscreen. But the important point is that once such a big freeze occurs, because of the reflectivity of the shiny surface of the planet it would be hard to have a big thaw. In fact, shiny ice is not really a good image of the appearance of snowball Earth. It would be so cold that tiny crystals of ice would form in the dry atmosphere and fall to the ground, where they would glitter like diamond dust.

The end of such a situation is almost certainly caused by a build-up of carbon dioxide in the air, warming the planet through the greenhouse effect. Under the conditions that exist on Earth today, greenhouse gases are emitted by volcanoes, but carbon dioxide dissolves in water which trickles over and through the rocks, where chemical reactions take carbon dioxide and use it in making rocks such as limestone. This weathering helps to keep things in balance. If the planet warms a little, there is more evaporation from the oceans, so there is more rain and more weathering, which draws carbon dioxide out of the air, so the greenhouse effect is reduced, and the world cools. When the world cools a little, the opposite happens. Human activities are in the process of upsetting this balance, but without us it has kept the temperature of the Earth stable within a relatively narrow range for millions of years, not least thanks to the hydrogen bonding which gives water its unusual properties.

During a snowball era, the Earth is so cold, with the equator as cold as the heart of Antarctica today, that there is essentially no weathering, so carbon dioxide can build up over a long period of time to the point where the temperature rises and a thaw sets in.

As the ice retreats, dark surface is revealed, and this absorbs solar warmth, raising temperatures further. But geologists estimate that the big thaw must take several million years to complete, after a snowball phase lasting tens of millions of years. Such thaws might be a major factor in our existence.

The snowball Earth event 2.5 billion years ago coincides, if that is the right word, with one of the most significant developments in the history of life on Earth. Just as the world was warming up again, huge amounts of oxygen were released into the air by the first organisms that evolved the ability to use carbon dioxide as food and release free oxygen through photosynthesis – single-celled creatures called cyanobacteria. This was a two-fold evolutionary advantage. To previous organisms oxygen was a poison that had to be locked away in harmless compounds, at the cost of energy. The new species could not only live with oxygen and save energy, but by releasing oxygen they poisoned all their rivals. The effect on the physical environment was equally dramatic. Free oxygen in the air and ocean reacted with iron compounds to form huge deposits of iron oxides seen in rocks around the world and known as 'banded iron formations'. As the world warmed out of the snowball state, it also rusted, thanks to the spread of photosynthesising life. Is it really a coincidence that a major evolutionary leap occurred just as life burst out of whatever niches it had survived in during the snowball Earth just as the world warmed? Probably not, although we may never know for sure.

We have a much better idea of how life exploded across the planet at the end of the most recent snowball phase, and much more compelling evidence that the thaw gave a boost to evolution.

Environmental stress can boost evolution by killing off success-ful species and allowing the survivors (assuming there are any) to adapt and evolve to take their place. The classic example is the death of the dinosaurs some 65 million years ago, which left empty ecological niches for the mammals to fill and adapt to, eventually producing ourselves. But the dinosaurs were only there themselves because of what happened about 600 million years ago.

This was before life had moved out of the sea and onto the land, so the species that survived the snowball must have been local-ised in rare warm places, perhaps associated with volcanic regions, where there was liquid water in puddles (reminiscent of Darwin's warm little pond). These survivors included bacteria, and larger single-celled organisms such as algae. Around the time that all this was happening, the first multicellular creatures, rather like sponges, evolved. This is just the sort of development that you might expect in an isolated warm little pond where new 'ideas' could get a start without much risk of being eaten by rivals. But it was just after the thaw that things got really interesting. Around 570 million years ago there was a proliferation of multicellular life so dramatic that it is used as the marker for the beginning of a new era of geological time, the Cambrian. It is often referred to as the 'Cambrian explosion'. Complex organisms of great variety evolved in the oceans at this time, as the evolutionary advantages of multicellular life allowed them to spread. Everything before the Cambrian is lumped together by geologists as the Precambrian – some 3.5 billion years of Earth history during which life was only represented by single-celled organisms. The few hundred million years since the Cambrian explosion contain almost everything that we multicelled creatures

233

regard as important, with life moving out of the sea onto the land and producing things as diverse as dinosaurs, oak trees, orchids and us. And it all began when the world warmed out of the most recent snowball state.

There are two messages to carry away from all this. The first is that hydrogen bonding is, from a human perspective, the most important pillar of science. It is responsible for the molecules of life, and, literally, for the water of life. The second message is that life is dramatically affected by things like snowball epochs. Without the snowball event 700 million to 600 million years ago and the subsequent Cambrian explosion we would not be here. And this is not the only bottleneck in the story of how life forms like us emerged on a planet like the Earth.

EPILOGUE

Bottlenecks: Maybe We Are Alone

What can the seven pillars of science tell us about our place in the Universe? The steps which led to the emergence of life on Earth are clear, and suggest that life is common in the Universe. But the possibility of other life forms like us, intelligent creatures with a technological civilisation, is much less clear. The technological qualification is important. By many criteria whales and dolphins are as intelligent as us, but they do not build radio telescopes and spaceships; if we are to make contact with other intelligent life in the Universe, it will be with creatures that do have that kind of technology. From now on, if I refer to intelligent life without qualification, I mean that kind of life. So why are we here, rather than 'only' dolphins and whales, butterflies and oak trees, or dinosaurs?

With only one example of a planet with this kind of intelligent life, it is unwise to generalise. But I shall do so anyway. One striking feature of our existence is how long it has taken for us to appear, both in terms of the age of the Universe and in terms of the age of the Earth. Our Solar System formed some 4.5 billion years ago, about 9 billion years after the Universe emerged from the Big Bang. There is a reason why it took so long for the Sun and its family of stars to form. The first stars were composed only of hydrogen and helium, and there were no heavy elements associated with them from which planets could form. Generations of stars had to run through their life cycles and spread heavier elements through interstellar space before even the small portion we find in the Solar System had built up in the cloud from which the Sun and planets formed. Using their understanding of the way stars evolve, and observations of our own island in space, the Milky Way, astronomers calculate that there is a 'Galactic Habitable Zone', or GHZ.

The Sun is part of a disc-shaped collection of stars, the Milky Way galaxy, about 100,000 light years across and 1,000 light years thick. Close to the centre of the Milky Way, there are many stars relatively close together, some of which explode as supernovas or kilonovas. This produces an abundance of heavy elements, from which planets can form around later generations of stars, but the radiation from these explosions is extremely harmful for life. Further out from the centre of the galaxy, there are fewer stars and less opportunity for heavy elements to build up. But in a ring around the Milky Way roughly 26,000 light years out from the centre, by about 5 billion years ago heavy elements had built up to the concentration we see in the Solar System, and stars like the Sun could form. We are close to the centre of this GHZ.

Once the Earth formed, as we have seen, life got started with almost indecent haste. But for more than 3 billion years this consisted only of single-celled organisms living in the sea. One inference from this is that even on Earth-like planets in our neck of the cosmic woods, that is the most likely kind of life to find. Was the emergence of multicellular life and the colonisation of the land inevitable? Or did it require a special event – a snowball Earth event – to trigger these developments?

Just as there is a Galactic Habitable Zone, so there is a Stellar Habitable Zone, or SHZ, defined as the region around a star where life forms like us can exist. The simplest rule of thumb is that this is the region where the temperature at the surface of a planet is between 0°C and 100°C, the range where, thanks to hydrogen bonding, liquid water can exist. The Earth is almost in the middle of the Sun's SHZ. As I pointed out earlier, the next planet in towards the Sun, Venus, although otherwise a prime candidate for the label 'Earth-like', is too hot. The next planet out from the Sun, Mars, is too cold today, although it may once have had a thick enough atmosphere for the greenhouse effect to bring its surface temperature into the critical range. Unfortunately, partly because it is a small planet with a weak gravitational pull, it has lost most of that atmosphere. This has led to astronomers coming up with another of their beloved acronyms, CHZ, for Continuously Habitable Zone. The Earth is near the middle of the Sun's CHZ, which extends only from about 5 per cent closer to the Sun than we are to 1 per cent farther out from the Sun than we are. The important point is that judging from how long it has taken for our kind of intelligent life to emerge on Earth, a planet does indeed need to be continuously habitable, at least for billions of years, to produce technological civilisation. If there ever was life on Mars, it never had the time to evolve into creatures like us. These are sobering considerations to set against the euphoria of headline stories about new discoveries of planets orbiting other stars.

The kind of orbits those planets are in also hints that there is something unusual about our Solar System. 'Our' planets go round the Sun in roughly circular orbits, and are spaced far enough apart that they do not have a great influence on one another. In other planetary systems, the orbits tend to be more elliptical – this rule holds particularly well for giant planets similar to Jupiter, the largest planet in our Solar System, which are easier to study. It is easy to understand how planets get into such orbits; this is the natural state for them to form in. It is hard to understand how the planets of our Solar System got themselves into neat circular orbits, and astronomers still argue over this. But the fact is that they are in such tidy orbits. You can imagine the chaos it would cause if Jupiter did have a significantly elliptical orbit, maybe moving in on each circuit of the Solar System as close to the Sun as the Earth is now, before swinging back out to the distance of Saturn today. Its gravitational pull would disrupt the orbits of any inner planets, which clearly would not be continuously habitable. By contrast, in its actual orbit Jupiter seems to have been a benevolent influence, helping to stabilise the Solar System and keeping the Earth habitable.

Jupiter, which has more than 300 times the mass of the Earth, has such a strong gravitational attraction that it has played a large part in the evolution of the Solar System. Early on, it was involved in shaking up the orbits of the bits of cosmic rubble left over from the formation of the planets that resulted in the Late Heavy Bombardment mentioned in Pillar Four. Once most of the debris had been cleared during this process, the rest was tugged by Jupiter into reasonably circular orbits between Mars and Jupiter, the asteroid belt, where it mostly remained. But there was also debris, in the form of ice-covered rocky material, left over in the outer part of the Solar System, beyond the orbits of the planets. This region is the source of comets which come in past the giant planets to the inner part of the Solar System and swing past the Sun, growing glowing tails as the icy stuff is evaporated by solar heating. Jupiter also captures many of these objects that might otherwise come in past the Earth, or even collide with our planet. This was spectacularly demonstrated in July 1994, when an incoming comet known as Shoemaker-Levy 9 was ripped apart by Jupiter's gravity and the fragments collided with the giant planet.

Even with Jupiter sweeping up much of this cosmic debris, some does still get through to the inner Solar System. The geological evidence reveals that the Earth has been hit by an object at least 10 km across once every hundred million years or so. This is about the size of the impact that struck our planet some 65 million years ago and caused a massive extinction of life on Earth, including the death of the dinosaurs. It has taken all that time for the shrew-like survivors of that disaster to have evolved our technological civilisation. Without Jupiter shielding us from such events, an impact like that would occur roughly every 10,000 years. There would be no chance to develop intelligence in such a short time, even if any complex life survived on land at all.

There are also threats to life on Earth from within our planet, not just from outside. About 250 million years ago, the Earth experienced a volcanic event (the word hardly seems strong enough!) which lasted for about a million years and spread lava to form a thick layer of rocks known as the Siberian Traps across what is now – you guessed – Siberia. This event and its associated impact on the

Impact of Shoemaker-Levy 9 with Jupiter

201010 Ltd/Science Photo Library

atmosphere and climate of the entire planet caused an extinction of life which killed about 90 per cent of all species around at the time, marking the end of the Permian era of geological time and the beginning of the Triassic.

There is also evidence of supervolcanoes on a smaller scale in the much more recent geological past. These include one that produced Lake Toba, in Indonesia, about 70,000 years ago. This was the largest known eruption of the past 25 million years. It spread a layer of ash roughly 15 cm deep over the entire Indian subcontinent, and while all that material and gases from the eruption were in the atmosphere it would have had a dramatic effect on climate. The environmental changes clearly had an impact on our ancestors. DNA evidence tells us that at just about the time of the Indonesian eruption the entire human population of the planet fell to about a thousand people. This is worth reiterating. The entire human population of the Earth, perhaps as little as a few hundred couples, only survived the catastrophe in an isolated pocket in East Africa. These numbers are so small that any species which existed today in such a precarious isolated population would be officially classified as endangered. We scraped through that bottleneck by the skin of our teeth.

You might feel reassured by the fact that this was indeed the largest eruption of the past 25 million years, and that we did survive. Surely there won't be another one soon? Think again. The entire region underneath Yellowstone Park in the United States is now known to be a supervolcano like this waiting to blow its top. Sooner or later it will erupt; we can only hope that it will be later rather than sooner.

The overall message is clear. The Earth is subject to repeated catastrophes, some from within, some from without – and I have not even mentioned events like 'ordinary' ice ages. On our planet, there has been time in one of the gaps between catastrophes for a technological civilisation to emerge, but only just. There is also some evidence that our planet is particularly favoured in this regard. The Solar System has been in the right part of the Milky Way at the right time to form planets like the Earth, and the unusual arrangement of the planets in our Solar System, especially the beneficial influence of Jupiter, has made the intervals between catastrophes unusually long. Does all this mean that even though life must be common in the Universe, intelligent life like us is rare and Bruno was wrong after all? You will have to make up your own mind about that, but my personal conclusion is that we are probably alone.

NOTES

1. See Andrew Weiner, https://www.jstor.org/stable/437245
2. *Six Easy Pieces*, Basic Books; 4th revised edition (7 April 2011).
3. From his *Autobiographical Notes*, edited by P.A. Schilpp, Open Court, Illinois, 1979.
4. *Science*, Volume 39, p. 791.

FURTHER READING

Easy Stuff

Richard Feynman, *Six Easy Pieces*, Basic Books, New York, revised edition, 2011

Steven Weinberg, *The First Three Minutes*, Basic Books, New York, revised edition, 1993

John Gribbin, *Stardust*, Penguin, London, 2009

Horace Freeland Judson, *The Eighth Day of Creation*, Cape, London, 1979

James Lovelock, *Gaia*, Oxford University Press, new edition, 2016

Not so easy stuff

Alexander Oparin, *The Origin of Life*, Dover, New York, revised edition, 1953

Linus Pauling, *The Nature of the Chemical Bond*, Oxford University Press, revised edition, 1960

Erwin Schrödinger, *What is Life?*, Cambridge University Press, 1944 (reprinted in 1967)

Hard stuff

Selected Genetic Papers of J.B.S. Haldane, Routledge, London, reprint of 1990 edition, 2015; also available on Kindle

Entertaining stuff

Fred Hoyle, *The Black Cloud*, Penguin Classics edition, 2010

EIGHT

IMPROBABLE

POSSIBILITIES

The Mystery of the Moon,
and Other Implausible
Scientific Truths

CONTENTS

For Steve Guest, who appreciates things like this!

LIST OF ILLUSTRATIONS

'When you have excluded the impossible, whatever remains, however improbable, must be the truth.'

The Adventure of the Beryl Coronet,
Arthur Conan Doyle

PREFACE

What Do We Know?

S cience deals with the unknown. My non-scientist friends sometimes offer sympathy when what is perceived as the 'failure' of
a scientific theory makes headline news. This happened recently
with the discovery that the expansion of the Universe is speeding
up, and that our simple Big Bang model needs modification. 'You
must be very disappointed,' they say, 'that your beautiful theory
is wrong.' Not at all! Good scientists are delighted when new evidence hints that new ideas are needed to explain what is going on
in the world. New ideas are the lifeblood of science, and if all our
theories were perfect descriptions of the world (by which I mean
everything there is, not just planet Earth), there would be nothing
left for scientists to do.

You might be surprised that there is anything much for science
to do at all. Given how much we already know about how the world
works, what is there left to discover? But a warning lesson from
history cautions against such complacency. Towards the end of the
nineteenth century, there was a widespread feeling among physicists that with Isaac Newton's theory of gravity and James Clerk
Maxwell's theory of electromagnetism they had all the tools they

needed to describe the world, and that no new fundamental discoveries remained to be made. In 1894 A.A. Michelson, an American physicist remembered for his work on measuring the speed of light, said:

> While it is never safe to affirm that the future of Physical Science has no marvels in store even more astonishing than those of the past, it seems probable that most of the grand underlying principles have been firmly established and that further advances are to be sought chiefly in the rigorous application of these principles to all the phenomena which come under our notice. It is here that the science of measurement shows its importance – where quantitative work is more to be desired than qualitative work. An eminent physicist remarked that the future truths of physical science are to be looked for in the sixth place of decimals.

It was just as well he put in the opening caveat, because hot on the heels of that remark came the discovery of radioactivity, the special and general theories of relativity, and quantum physics. Definitely marvels even more astonishing than those of the past. Scientists have learned never to say that all that remains is to dot the i's and cross the t's of their favoured theories.

How can there be more to be discovered when so much is already known? An analogy may help. Pretend that everything we know about the world is represented by the area inside a small circle drawn on a large, flat piece of paper. Everything we know is inside the circle, everything we don't know is outside. As we discover more about how the world works, the circle gets bigger. But as it does so, the circumference of the circle, the boundary between what we know and what we don't know, also gets bigger. As the Lovin' Spoonful song 'She is Still a Mystery' puts it, 'the more I see, the more I see there is to see'. There will be plenty of work for scientists in the foreseeable future. And that work proceeds by setting up hypotheses (or guesses) about how the world works, then carrying out experiments or making observations to eliminate the incorrect guesses.

Are relativists delighted when a new observation of the Universe confirms, as the headline writers like to put it, that 'Einstein Was Right'? Only up to a point. What would be really exciting for them would be an observation which showed that the general theory of relativity is good as far as it goes, but that it may not be right everywhere and all the time. That is why such experiments are carried out. Not to 'prove Einstein was right' but in the hope of finding out the conditions, or places, in the Universe where Einstein's theory might be wrong.

So in spite of what popular media may tell you, good scientists do not carry out experiments in order to prove their pet theory is right.* They carry out experiments in order to find where the theory fails, which tells them where new discoveries can be made (and, if you care about such things, where Nobel Prizes might be won).

As Richard Feynman famously pointed out:

If it disagrees with experiment, it is wrong. In that simple statement is the key to science. It does not make any difference how

---

* There are, of course, bad scientists who do just that, but they have no place here.

beautiful your guess is, it does not make any difference how smart you are, who made the guess, or what his name is – if it disagrees with experiment, it is wrong.

This is the scientific equivalent of Conan Doyle's dictum. It is by experiment (or observation) that scientists eliminate the impossible. Thomas Henry Huxley called this 'The great tragedy of science – the slaying of a beautiful hypothesis by an ugly fact.'

But a good scientist doesn't go quite as far as Doyle does. Once you have eliminated the impossible, whatever is left is certainly possible, in the light of present knowledge, but may not be the ultimate truth. It may yet, in its turn, be slain by an ugly fact. It is with that in mind that we should turn our attention to some of the improbable (in the light of present knowledge) truths of science.

IMPROBABILITY

The Mystery of the Moon

A total eclipse of the Sun is one of the most spectacular and beautiful sights visible from the surface of the Earth. It is so spectacular because the Moon and Sun look the same size to us. So when the Moon passes in front of the Sun, it can exactly cover the bright solar disc, plunging the region affected by the eclipse into darkness, but allowing the glowing outer layer of the Sun, its corona, to become visible like a glorious halo. But why are we lucky enough to see this sight? Why are the apparent sizes of the Sun and Moon just right to produce it? The question is more profound than it seems at first, because the coincidence has not always held. Our human civilisation exists at a rare moment of astronomical time when the Moon is perfectly placed to make this kind of eclipse. In the not too distant geological past, it was too close to Earth and would have blotted out the corona as well; in the astronomical future it will be too far away and will look like a small dark blob passing across the solar disc. Improbably, it is 'just right' just at the time we are here to notice it.

A solar eclipse
Science Photo Library

But the effect only happens at all because the Moon is so large. As a fraction of the size of its parent planet (Earth), it is by far the largest moon in the Solar System. Indeed, many astronomers think that the Earth–Moon system should better be regarded as a double planet than as a planet plus a moon. And that is all down to the way the double planet formed.

The Sun and Solar System formed when a cloud of gas and dust in space collapsed under the pull of its own gravity. Most of the material went in to the central star, the Sun. Some of the dust, and icy particles, was left in a disc around the star, and particles of that dust collided and stuck together until some were big enough to tug other particles towards them by gravity, so that bigger and bigger objects built up. This eventually made the planets, but some material was left over to make smaller objects, asteroids and comets. The late stages of this process were far from gentle, as proto-planets were bombarded with debris as they swept their orbits around the Sun clear. Just a hint of what this bombardment was like can be gleaned from the battered face of the Moon; but this tells less than the full story, because the Moon itself only formed after most of the process of planet building had taken place.

It is straightforward to account for the moons that we see orbiting around other planets in the Solar System, such as Mars, Jupiter and Saturn. The moons of Mars are clearly small pieces of debris – asteroids – left over from the planet-building process and captured by Mars. The moons of giant planets like Jupiter and Saturn are much bigger than asteroids – but the giant planets are much bigger than Mars. Their families of moons formed around the parent planets in the same way that the planets formed around the Sun, making miniature 'solar systems'. But the Moon is 25 per cent as big as the Earth, in terms of its diameter, and clearly formed in a different way. The best explanation is that within a few million years of the Earth forming, the planet was involved in a collision with another young planet, an object the size of Mars, which struck it a glancing blow. In the heat generated by this violent event, the incoming object would have been destroyed, and the proto-Earth's newly formed crust would have melted. The heavy metallic core of the incomer would have sunk to the centre of the Earth, mixing with Earth's own metallic heart to make a planet with a very dense core and a relatively thin crust. The crust would be thin because molten material from the impact, a mixture of stuff from the proto-Earth and the incomer – graphically referred to by astronomers as the Big Splash – would have been flung off into space, some escaping entirely but some staying to form a ring around the Earth from which the Moon coalesced. It is easy to remember how long this process took; computer simulations tell us that something resembling the Moon would have formed within a present-day month of the impact. Dating of lunar rock samples tells us that all this happened about 4.4 billion years ago. Among other things, the impact set the Earth spinning rapidly on its axis, and knocked it out of the vertical, causing the tilt which is responsible for the cycle of the seasons.

All of this explains many oddities about the Earth. The planet Venus, just sunward of the Earth, is roughly the same size as the Earth, but has a thick crust, a small metallic core, and as a result a negligible magnetic field. It rotates only once every 243 of our days. The Earth has a thin crust, a large metallic core that is responsible for a strong magnetic field, relatively rapid rotation, and a large Moon. These features go together like a hand in a glove. Our planet is the odd one out in the Solar System, produced by a highly improbable sequence of events, all linked to the Moon. And the consequences of those events are far-reaching.

Take the thinness of the crust. It might not sound like a big deal, but it is. The crust is so thin that it can crack like an eggshell, with the pieces of the shell being moved about by convection currents in the fluid layers beneath, in the process known as plate tectonics. Thanks to the thinness of the crust, around the edges of these pieces of shell (the plates) there is constant volcanic activity, releasing gases like carbon dioxide and water vapour into the atmosphere. Where the crust is cracked, usually under the oceans, new crust can be made as molten material wells up and sets, spreading out on either side of the crack, pushing the plates away on each side. But the Earth does not get any bigger, because in other parts of the world, especially along the edges of some continents, crust is being pushed down into the interior. This carries carbonates and water back down where they get fed into volcanoes and are released into the air again in an endless cycle.

But the cycle does not run at a constant speed. The process which takes gases like carbon dioxide out of the atmosphere is called weathering. Carbon dioxide dissolves in water, and then reacts with minerals in the rocks to make calcium carbonate (limestone). Carbon dioxide in the atmosphere is, of course, a greenhouse gas – it traps heat and keeps the surface of the Earth warmer than it would otherwise be. As it happens, weathering proceeds faster when the world is warmer, so that tends to draw carbon dioxide out of the air efficiently, allowing the planet to cool. But when it cools, weathering is less efficient, and carbon dioxide builds up in the air again. The world warms, and the weathering process speeds up, drawing more carbon dioxide out of the air. There is a negative feedback which, thanks to plate tectonics, helps to keep the temperature at the surface of the Earth in the range where liquid water can exist (although, unfortunately, these natural processes are too slow to compensate for the buildup of carbon dioxide now being caused by human activities quickly enough to save us from the consequence of our own folly). Without this process – without the thin crust produced by the impact that made the Moon – the Earth would probably have become a scorching desert with a thick carbon dioxide atmosphere, like our neighbour Venus.

This isn't the only thing we have to thank the Moon for. Analysis of seismic waves produced by earthquakes and travelling through the interior of our planet shows just how large the central core is. It is a solid lump of iron and nickel with a diameter of about 2,400 km, the top of which is about 5,200 km below the surface of the Earth. But it is surrounded by a layer of liquid material, extending a further 2,500 km upward, roughly halfway to the surface of the Earth from the top of the inner core. Together, the inner and outer core contain a third of the mass of our planet, part of it donated by the impacting object which produced the Moon. It is the outer core that is important to us, and to all life on Earth. The temperature in this iron–nickel liquid layer is about 5,000°C, only a little less than the temperature at the surface of the Sun, maintained by the radioactive decay of elements such as thorium and uranium, left over from the formation of the Solar System. Swirling currents in this layer generate the magnetic field of the Earth.

The Earth's magnetic field is literally a force field, which protects our planet from a major threat from space. The Sun produces a blast of electrically charged particles, blandly called the 'solar wind', which reaches out from its source across space and past the Earth and the other planets. These particles travel at speeds of several hundred kilometres per second most of the time, and up to 1,500 kilometres per second during outbursts known as solar storms. Without the shielding effect of the magnetic field which forms a protective layer around the Earth, these 'solar cosmic rays', essentially the same as the particle radiation from a nuclear bomb, could strip away the outer layers of the atmosphere and penetrate to the ground where they would cause considerable damage to life, possibly even sterilising the land surface of the planet.

The region around the Earth that is protected by the magnetic field is called the magnetosphere, but 'sphere' is actually the wrong term, because the solar wind is so powerful that it squashes the magnetic field on the side facing the Sun, while on the other side of the Earth the magnetic field is stretched out in a long tail, making an overall shape like a cosmic tadpole. On the side facing the Sun, the boundary between the magnetic field and the solar wind (the hull of Spaceship Earth) lies about 64,000 km above the surface of the Earth; on the side away from the Sun, it stretches out almost exactly as far as the distance to the Moon. And at the north and south magnetic poles, a small proportion of the particles of the solar wind leak in to the upper part of the atmosphere of the Earth. Most of the time, the only effect this has is to produce the beautiful displays known as the northern and southern lights. But during solar storms the effects at high latitudes can be damaging to anything that uses electricity. They disrupt communications, affect power lines, and cause blackouts in places such as Canada. If the magnetosphere suddenly failed, this would happen all over the Earth.

It is a sobering fact that there is geological evidence that just such events have occurred in the past, with the magnetic field fading away suddenly (by the standards of the geological timescale) then rebuilding, either in the same sense as before or with north and south magnetic poles reversed. The evidence comes from the magnetic record left in some kinds of rock as they solidify after volcanic eruptions. As the rock sets, the magnetic field gets frozen in to it, forming a permanent magnet preserving the direction of north and south at the time. The rocks can be dated by various techniques to show when the magnetic field gradually disappeared. And the fossil record of life on Earth shows that when the magnetic field is weak, many species of life on Earth go extinct, although creatures living in the oceans are not affected. The natural conclusion is that land dwellers were zapped by radiation from space, while sea dwellers were protected by layers of water. But even if this explanation is wrong, there is no escaping the evidence that land dwellers die out when the field is weak. The not-so-cheery news is that over recent decades the Earth's magnetic field has been weakening at a rate somewhere between about 5 per cent per century and 5 per cent per decade. If this continues, it could disappear some time between about 2,000 years and 200 years from now.

Partly because the Earth gained some of the heavy elements that make up the core during the Moon-forming impact, while lighter material splashed out into space, even though the diameter of the Moon is a quarter that of the Earth its mass is only one eightieth of the mass of the Earth. Even in those terms, however, this still makes it the largest moon in proportion to its planet in the Solar System.* Because of this, the gravitational influence of the Moon on the Earth has been a major factor on our planet ever since the Big Splash. The most obvious manifestation of this influence today is in the tides, but these are just a feeble ripple in the sea compared with what they used to be.

Computer simulations tell us that when the Moon first formed it was orbiting only about 25,000 km above the Earth, compared with an average distance today of a bit more than 384,000 km. This would have raised enormous tides not just in any oceans that existed but in the 'solid' Earth as well, stretching and squeezing the rocks over a range of about a kilometre in a regular rhythm. At first, the heat generated by this process would have kept the rocks molten even after the Big Splash, so the tides actually involved oceans of lava. But the energy of that process came from the orbital energy of the Moon, and as the energy was lost, it made the Moon weaken its grip and move outwards while the tides got smaller. A solid crust had formed by about a million years after the collision that gave birth to the Moon.

Thanks to the impact, the Earth was also spinning rapidly then, so that a day was about five hours long when the Moon was young. Today, we have tides about a metre high roughly twice a day, every twelve hours or so, with variations caused by the local geography of coastlines. Just (a million years or so) after the Moon formed, there

---

* If you want to argue that Pluto is a planet and has a proportionately very large moon, Charon, my response would be that Pluto–Charon is a double planet.

were tides several kilometres high about every two-and-a-half hours. Life emerged from the sea and moved on to the land about 500 million years ago, and even a hundred million years later, 400 million years ago, in a memorable numerical coincidence there were about 400 days in the year, because the Earth was still spinning about 10 per cent faster than it does today, each day then being only a little more than 21 hours long. But over the billions of years since the Moon formed, one thing has stayed reasonably constant – the tilt of the Earth. And once again we have the Moon to thank for that.

Spinning objects that have a tilt wobble, as anyone who has played with a child's top knows. But there is more than one kind of wobble. The Earth leans over in space by about 23.4 degrees from a line at right angles to the plane of the Earth's orbit around the Sun. As I have mentioned, this tilt was produced when the young Earth was struck by a Mars-sized object in the collision that created the Moon. Over the course of a year the tilt always points in the same direction, so as the Earth moves around the Sun sometimes it leans towards the Sun, and sometimes away from the Sun. This is not a wobble, as you can visualise if you pretend that it is the Sun moving around a stationary Earth. The tilt causes the cycle of the seasons – when one hemisphere is leaning towards the Sun it is summer there and winter in the opposite hemisphere, and when one hemisphere is leaning away from the Sun it is winter there and summer in the opposite hemisphere.

I was careful to say that the tilt always points in the same direction 'over the course of a year', because it does actually change slightly in a regular way over tens of thousands of years. This really is a wobble, and has profound and improbable implications for life on Earth, which I discuss in Improbability Eight. But here I am more interested in why the wobble isn't bigger. This is, of course, thanks to the stabilising influence of the Moon's gravity. The planets (and moons) of the Solar System all tug on one another by gravity, producing an influence which changes as the planets move round their orbits, and smaller planets such as Earth and Mars are particularly susceptible to the combined influences of the largest objects in the Solar System, the Sun and Jupiter. If a planet like Earth or Mars was the only planet orbiting the Sun, it would go on its way without wobbling. But, improbable though it may seem, even small gravitational nudges from the Sun and Jupiter can induce big wobbles, through the process known as chaos, which features in Improbability Six.

Computer modelling tells us that on Mars, which has no large moon, the tilt can change suddenly by at least 45 degrees, and more slowly up to about 60 degrees, where 'suddenly' means over the course of about 100,000 years. We don't have to rely solely on the computer modelling, though, because the surface features of Mars have now been studied by orbiting spaceprobes in enough detail to confirm that over geological time this kind of change has indeed occurred. This gives us confidence in the predictions of the same kind of modelling applied to our own planet, which tell us that without the presence of the Moon the Earth could go from being nearly upright in its orbit to nearly flat, with a 'tilt' of almost 90 degrees, over as little as 100,000 years. The implications would be profound. With one pole pointing towards the Sun, that hemisphere would experience searing summer during which the Sun never set, while the opposite hemisphere froze over as the Sun failed to rise. Six months later the situation would be reversed. And the tropical regions would be in permanent twilight and never thaw at all. It is solely thanks to the presence of the Moon that nothing like this has happened since life emerged onto land (as we know from the fossil record), and probably for much longer than that (as we infer from the computer modelling).

It is, of course, too good to last. As the Moon slowly but steadily retreats from the Earth, its stabilising influence will get less and less. The Moon has been with the Earth and exercising that influence for a bit more than 4 billion years, and is now moving outwards at a rate of about 4 cm per year. The simulations tell us that in about 2 billion years from now its stabilising influence will be too weak to prevent Jupiter-induced toppling of the Earth. Which brings me back to the improbability I started with. The Sun is about 400 times bigger than the Moon, but it is also about 400 times further away from us than the Moon is. In the past, the Moon would have looked much bigger, and easily blotted out the Sun during an eclipse. But during the era of the dinosaurs there was nobody around to notice. In the not too distant future (long before that wobble occurs), a ring of sunlight will be visible around the edge of the Moon even during an eclipse. There may or may not be anyone around then to notice. How curious that intelligent beings who notice things like that should be around just at the moment of geological time that it is there to be noticed. Especially because we are only here at all because of the Moon and its influence on the Earth. Highly improbable – but not impossible, as the fact that I am telling you about it testifies.

The Universe Had a Beginning, and We Know When it Was

The idea of the Big Bang origin of the Universe is now so familiar that it has become a cliché. It has even been used in the UK to refer to the sudden deregulation of the financial markets under Margaret Thatcher. But the idea of a beginning to the Universe is so improbable that scientists never even considered it until about a hundred years ago, and it did not become firmly established until about 50 years ago.

To the ancients, looking out at the night sky, the Universe seemed eternal and unchanging. Right up until the 1920s, what we now know to be the Milky Way galaxy, an island of a few hundred billion stars, was thought to be the entire Universe, in which individual stars might be born, live and die but the overall appearance always remained the same, like a forest in which individual trees live and die without changing the overall appearance. The idea of an unchanging Universe was so ingrained that even Albert Einstein, usually willing to entertain new ideas, accepted it without question.

When he applied the equations of his general theory of relativity to describe the behaviour of the entire Universe (all of space and time), he found that the mathematics said that the Universe could not be static, but must be either expanding or contracting. This seemed so improbable to him that he added an extra factor to the equations, called the cosmological constant, to hold everything still.

In the early 1920s, improved telescopes and photographic techniques led to the discovery that our Milky Way is not the entire Universe, but just one island of stars among many scattered through vast regions of space. But at first this still seemed to fit the idea of a static Universe, albeit on a bigger scale. Then, at the end of the 1920s, Georges Lemaître and Edwin Hubble independently discovered (from the famous redshift effect) that the galaxies (strictly speaking, clusters of galaxies) are moving apart from one another – that the Universe is expanding. This was interpreted as caused by a stretching of space, exactly in line with Einstein's equations *without* the cosmological constant. He later described introducing the constant as 'the biggest blunder' of his career. Improbable as it had seemed, the Universe really was expanding.

But did that mean it had a beginning? Not necessarily. Some cosmologists argued that because galaxies are moving apart today, then long ago they were packed together in one lump of stuff, a kind of cosmic egg, that exploded outwards. But another school of thought held that as the galaxies moved apart the empty space between them got filled in by new galaxies being created out of primordial energy. This continual creation of matter seemed no more extravagantly improbable than the idea that all the matter in the Universe had been created in one go in some kind of cosmic egg. The continual creation would allow for the Universe to be eternal and unchanging in its overall appearance even though it was expanding. This 'Steady State' model was championed by Fred Hoyle, who coined the term Big Bang in a BBC radio broadcast to highlight the distinction between the two ideas.* To Hoyle and other proponents of the Steady State idea, the idea of a definite beginning seemed too improbable to take seriously. They weren't the only ones, as it has turned out. Soon after the discovery that the Universe is expanding, in 1931 Albert Einstein turned his attention to the implications. He drafted a paper in which he came up with exactly the same idea that Hoyle developed a decade and a half later, and wrote:

> If one considers a physically bounded volume, particles of matter will be continually leaving it. For the density to remain constant, new particles of matter must be continually formed in the volume from space.

But he got distracted by other work and never finished the paper for publication. It languished in the archives until Cormac O'Raifeartaigh and Brendan McCann, of the Waterford Institute of Technology, came across it eight decades later and had it translated and published in English in 2014. So if anyone asks you who was the first person to come up with the Steady State/Continual Creation idea, the answer is Einstein!

....

* It is often said that he meant this as a term of derision. He told me, however, that he was simply looking for a snappy expression as a contrast with the expression Steady State.

The debate between the two camps raged through the 1950s and into the 1960s. What was needed was a test to distinguish between the predictions of the two ideas, and one was dreamed up, but not initially carried out, roughly at the same time the Steady State idea was first being developed. It essentially depended on the idea that if the Universe were smaller in the past, with everything squashed together more tightly, then it must also have been hotter then, in the same way that the air in a bicycle pump gets hotter when it is compressed. From basic physical principles, two young American researchers, Ralph Alpher and Robert Herman, calculated how hot the Universe must have been when it was as dense as the nucleus of an atom, at the time of the Big Bang,* and how hot the leftover radiation from the Big Bang must be today. In 1948, they published their conclusion that 'the temperature in the universe at the present time is found to be about 5 K', which corresponds to minus 268 degrees Celsius. The idea was promoted by their senior colleague George Gamow, whose name is often linked with the calculation, although he did not carry it out.

The prediction was largely forgotten in the 1950s, but early in the 1960s – less than 60 years ago – Arno Penzias and Robert Wilson, working at a radio telescope owned by the Bell Laboratories, unexpectedly discovered that the Universe is filled with a sea of micro-wave radiation with a temperature of about 3 K, later determined

---

* We call this density the Big Bang because we understand the physics of matter at such densities and everything since very thoroughly. How this hot, dense fireball arose (what happened before the Big Bang) is more speculative, but I shall look at the best explanation shortly.

more precisely to be close to 2.7 K. This was a double surprise because not only were they unaware of the work of Alpher and Herman, they were both supporters of the Steady State idea. But they had accidentally ruled this out as impossible. It was quickly realised that this must be the radiation predicted by Alpher and Herman, and that there really had been a Big Bang, improbable though that still seemed to many astronomers. So *when* had the Big Bang happened? How old is the Universe?

The original technique for working out the time since the Big Bang depended on measuring the speed with which galaxies seem to be moving away from us (easy) and measuring how far away they are (hard) so astronomers could work backwards to find out when everything was together in one place. This is trivial arithmetic – if a car travelling along a straight highway at 60 miles an hour is 30 miles away from its starting point, how long ago did it start? The 'speed' is measured directly from the redshift, which is a stretching of light caused by the expansion of the Universe. It is not a Doppler effect, despite what some accounts may tell you, because it is not measuring a speed through space, but the speed with which space itself is expanding, carrying galaxies along for the ride. The distances to galaxies are hard to measure, and this depends on knowing (or guessing) things like how bright galaxies are so the distance can be estimated from how dim they look to us – like measuring the distance to the end of the street by measuring how faint the light from a street lamp is. The relationship between speed and distance is described by a number called the Hubble constant, H. The bigger H is, the faster the Universe is expanding and the less time there has been since the Big Bang.

The radio telescope that Penzias and Wilson used to discover the CMB
NASA

In the 1960s, when I started my career (such as it was) as an astronomer, the difficulty of working out distances to galaxies (the distance scale) meant that the best astronomers could do was to say that H must be somewhere between 50 and 100, and was probably about 75. For a value of 100, the age of the Universe would be a bit less than 9 billion years, while a value of 50 implies that it is twice as old, about 18 billion years. Quite separately, however, astrophysicists had been developing techniques to estimate the ages of stars, and were finding that the oldest known stars were significantly older than 9 billion years, which made the largest value of H suggested by cosmologists impossible.

Over the next few decades, improved measurements based on the traditional technique, culminating in studies made with the Hubble Space Telescope, pinned down the value of the Hubble constant more and more accurately, culminating in a value of 72 ± 8 (that is, between 64 and 80), reported in 2001. But meanwhile, a completely different technique, using observations of the cosmic microwave background radiation, had also got a handle on the Hubble constant.

When the background radiation was first identified and astronomers took its temperature they found that it was exactly the same everywhere they looked – the temperature of the sky is the same in all directions. This matches the simplest predictions of cosmological calculations of the Big Bang, and among other things confirms that we do not live in a special place in the Universe, since this kind of pattern (or lack of pattern) would look the same from anywhere in the Universe. But as their measurements improved, and still no pattern was detected, this began to raise a nagging worry in the minds of cosmologists. Those simplest cosmological calculations actually described the behaviour of uniformly expanding spacetime without any matter in it. The real Universe contains galaxies of stars, and these must have grown out of irregularities that were present way back when the Universe was a hot fireball and the background radiation was much more intense. At that time, the fireball would have contained a sea of electrically charged particles, protons and electrons, interacting with the electromagnetic radiation of the fireball so that its temperature at any one spot depended on the density of the matter at that location. Then, as the Universe cooled to a temperature of a few thousand degrees (roughly the same as the temperature at the surface of the Sun today, but everywhere in the Universe) the protons and electrons got locked up in electrically neutral atoms, and the radiation 'decoupled'. It still carried the imprint of those primordial fluctuations but no longer interacted strongly with matter as it cooled all the way down to 2.7 K.

These primordial irregularities should have left an imprint on the temperature from different parts of the sky which could be detected today, if we had sensitive enough instruments to measure it. But they would have to be *very* sensitive. Working backwards from our measurements of the sizes of galaxies and clusters of galaxies today, astronomers were able to calculate how uneven the Universe was just after the Big Bang, at the time of decoupling. This meant they knew the size of the fluctuations in the temperature of the background radiation from place to place at that time, and could then work forwards again to calculate how big the differences in temperature from one part of the sky to another must be today. The answer turned out to be one part in 100,000. Which, for an average temperature of 2.7 K, meant that the instruments had to detect fluctuations of 0.00003 K, 30 millionths of a degree.

Improbably, measurements this accurate were made using a satellite called COBE (COsmic Background Explorer) launched by NASA in November 1989. Almost immediately, the instruments on COBE were able to measure the average temperature of the background radiation more accurately than ever before, and the result, 2.725 K, was presented at the January 1990 meeting of the American Astronomical Society. But that was just the beginning. Over the course of more than a year, the instruments on board COBE scanned the entire sky, using three separate detectors. They took 70 million separate temperature measurements, which the team responsible for the mission then had to analyse, subtracting out the average temperature to produce a map which showed the tiny differences in temperature from one spot on the sky to another. The map was completed in 1992, and it revealed ripples in the background radiation, with hot spots on the sky 30 millionths of a degree warmer than average, and cold spots on the sky 30 millionths of a degree below average. But the average, remember, is just 2.725 K, so 'hot' and 'cold' are only relative terms. Traces of the irregularities which, under the influence of gravity, had grown to become galaxies had been found, providing more evidence in support of the Big Bang idea. But this alone did not provide an accurate independent measurement of the age of the Universe.

Inspired by the success of COBE, astronomers set out to squeeze even more information out of the background radiation. This involved searching for traces of so-called 'baryonic acoustic oscillations', which is jargon for sound waves – sound waves in the early universe that have left an imprint on the background radiation that might be detectable today.* These secondary ripples are smaller and harder to detect than the ones found by COBE, but they carry additional information which makes the effort of detecting them worthwhile. The exact pattern made by these ripples depends on a balance between the gravitational forces pulling huge clouds of gas together, and the influence of fast-moving particles of light (photons) of the background radiation during a short period of time, about 100,000 years, just after decoupling. This tended to smooth out irregularities. Some wavelengths of the sound get bigger during this interval while others are smoothed away. And this all happens in the expanding Universe, which both cools the photons (making them less energetic and reducing their influence) and stretches the acoustic waves, so that the speed with which the Universe is expanding, which depends on the Hubble constant, also comes in to consideration.

The result is a messy mixture of wavelengths, but astronomers are used to dealing with such mixtures and have a powerful tool, called power spectrum analysis, which can pick out the individual wavelengths that contribute to the messy overall picture. This is a bit like analysing the sound made by a symphony orchestra in full voice and working out what individual contributions are being made by the violins, the flutes, the percussion and all the other instruments. Another analogy would be analysing the sound that comes out of a church organ to work out the lengths of the organ pipes and other details of the structure of the instrument.

....

* 'Baryon' is just the generic term for things like atoms, anything mostly made, in terms of mass, of protons and neutrons.

Power spectrum analysis of the pattern of tiny temperature fluctuations in the background radiation across the sky produces a wiggly graph, called (logically enough) the power spectrum, with a large peak on the left and a series of smaller and smaller wiggles tailing off to the right. The relative heights of the peaks in the graph provide a lot of data (not just about baryon acoustic oscillations, although they are particularly important) which reveals information about many features of the Universe, including the speed with which it is expanding and therefore the value of H and the age of the Universe. The key thing to remember is that this measurement is entirely independent of the traditional technique based on measuring the distances to galaxies.

The power spectrum of the microwave background was studied by two satellites in the early decades of the twenty-first century. First came NASA's Wilkinson Microwave Anisotropy Probe (known as WMAP and pronounced double-you-map), launched in the summer of 2001. The detectors on board WMAP were 45 times more sensitive than those on board COBE, and they could measure the temperature from individual patches of sky roughly one-fifth of a degree across, a third of the size of the full Moon as seen from Earth today. It operated until 2010, when it was moved to a parking orbit out of the way of any future satellites and switched off. Among the wealth of data it obtained (some of which comes into the story of Improbability Three) it initially measured the Hubble constant as 72 ± 5, corresponding to an age of the Universe (the time since the Big Bang) of 13.4 ± 0.3 billion years. As time passed and more data were gathered, including observations made by instruments carried on high-altitude balloons, this estimate of the age of the Universe was made even more precise and pushed up slightly to 13.772 ± 0.059 billion years. But by the time WMAP was being switched off and parked, another satellite, the European Space Agency's Planck probe, was picking up the baton.

Planck was launched in May 2009. With a sensitivity three times better than WMAP its instruments could measure differences in temperature from one spot on the sky to another as small as a millionth of a degree, while the size of the spots it measured was only one-twentieth of a degree across. Stop and think about that. It is easy to bandy about terms like a millionth of a degree. But in the second decade of the twenty-first century astronomers could look at a patch of sky one sixth of the apparent size of the full Moon, and tell you its temperature relative to its neighbours *to an accuracy of one millionth of a degree.* If that doesn't boggle your mind, nothing will.

Planck operated until 2013, when it followed WMAP into a parking orbit. It was in March that year, just before the satellite was retired, that the first detailed results from Planck were announced, indicating an age of the Universe of 13.819 billion years. A couple of years later, with more data analysed, the Planck team revised their estimate to 13.799 ± 0.021 billion years and a value of 67.74 ± 0.5 for the Hubble constant. This is not only very close to the number found by WMAP, but both satellites give values right in the range of 72 ± 8, reported in 2001 by the people using the traditional technique. In less than 60 years, astronomers had gone from arguing about whether the value of H is 50 or 100, corresponding to ages of 18 billion or 9 billion years, to quibbling about a difference in the second decimal place! That difference between the measurements made by WMAP and Planck amounts to no more than 100 million years out of roughly 14 *billion* years – less than 1 per cent.

This is one of the greatest – and most improbable – achievements of science. Any scientist, let alone a lay person, of any previous generation would have been dumbfounded to learn that we know the age of the Universe to within 1 per cent, and that it is 13.8 billion years, give or take a hundred million. But even this is not the end of the story. There is some icing to put on the cake.

I mentioned earlier that there was concern back in the 1960s (and even a little later) when some estimates of the age of the Universe came out lower than estimates of the ages of some stars. There is no longer any need for that concern. While cosmologists were busy refining their estimates of the age of the Universe, astrophysicists were equally busy refining their measurements of the ages of stars. With gratifying results.

There are now several different ways to measure the ages of stars, but I will briefly mention just two, to highlight how well astronomers now understand the Universe. The first depends on the discovery, made early in the twentieth century, of how the temperature of a star (which is linked to its colour) is related to its brightness. On a graph plotting brightness on the vertical axis and temperature going down from right to left along the horizontal axis, most stars lie on a line running from top left (hot and bright) to bottom right (cool and dim). The apparent brightness of a star depends on its distance, so these brightnesses are calculated from how a star would look at a distance of 32.5 light years (10 parsecs, in the units astronomers prefer). That in turn depends on knowing the distances to stars, which is why this relationship was not discovered sooner (just how astronomers measure distances to stars is a whole other story). The graph is called the Hertzsprung-Russell (or H-R) diagram, after the two astronomers who each independently discovered the relationship. Because most bright stars are on the top left to bottom right line, it is called the main sequence. But there are smaller numbers of stars below the line to the left (dim but hot) and above the line to the right (bright but cool).

The position of a star on the main sequence depends only on its mass. The more massive a star is, the faster it has to burn the nuclear fuel in its heart to hold itself up against its own weight, so it releases more energy and is very bright. But as its nuclear fuel is being exhausted, the outer regions of the star swell up, so all the heat is going across a bigger area, which makes the surface cool. The star becomes a red giant, in the upper right of the H-R diagram. When all its fuel is used up, it becomes a stellar cinder (a white dwarf), shrinking down and appearing in the bottom left of the diagram. But the key point is that the time in a star's life when this happens depends on its mass. So for a group of stars all the same age, the main sequence steadily gets shorter, as if the line was being rubbed out starting at the top left. In the memorable expression, 'big stars live fast and die young'. The relationship between mass and temperature of main sequence stars is very well understood, because whatever is holding the star up must produce exactly enough heat to stop it collapsing. Too much heat and the star would explode; too little and it would shrink. So the point where the main sequence ends tells us the mass of the oldest stars in that group, and that in turn tells us their age, because we know how long it takes for stars of different masses to use up their nuclear fuel (essentially by converting hydrogen into helium).

Putting all this together, if we have a group of stars all the same age, *and* we can measure the distance to them, *and* we can identify the top end of their main sequence, then we can work out the age of the group of stars. Fortunately, there are such groups, called globular clusters. These are spheres of stars which contain hundreds of thousands of individual stars. The stars in each cluster all formed together not long after the Big Bang, in the outer regions of the primordial cloud of gas from which our galaxy grew (similar clusters are seen around other galaxies). Unfortunately, working out the distances to them is very hard. But at least standard physics does tell us the ages of stars with different masses. With all the difficulties involved, even by the middle of the 1990s all the astrophysicists could say about the ages of globular clusters was that they must be between 12 and 18 billion years old. At least this was in the same ball park as cosmological estimates of the age of the Universe, but frustratingly vague. But then came the ESA satellite Hipparcos.

Hipparcos was launched in 1989, and made precision measurements of the distances to almost 120,000 stars, using the technique of parallax, the apparent shift in position of stars on the sky when they are measured from opposite sides of the Earth's orbit. The team described the accuracy of the measurements as equivalent to using a telescope on top of the Eiffel Tower to measure the size of a golf ball on top of the Empire State Building. Combining Hipparcos data with all other ways of measuring the ages of globular clusters, by the end of the 1990s Brian Chaboyer and other members of the Hipparcos team came up with a best estimate of the ages of the oldest globular clusters of 12.6 billion years, with an uncertainty of roughly plus or minus a billion. Data from a later satellite, Gaia, launched at the end of 2013, suggest a slightly higher figure, but still less than 13.8 billion years.

The second technique I want to mention is very, very simple in practice but horrendously difficult for the observers to carry out. It depends on the way radioactive elements 'decay' to produce a mixture of different elements. This process is very well understood from studies here on Earth, and is widely used for estimating the ages of rocks. Uranium-238 is one such useful radioactive element, and in any sample of U-238 half of the atoms decay in 4.5 billion years (roughly the same as the age of the Earth), half of the remainder decay in the next 4.5 billion years, and so on. The age of anything containing U-238 can be worked out by measuring how much U-238 it contains now and comparing this with the quantities of the various 'decay products' which tell us how much it started out with. 'All' the astronomers had to do was find a star with uranium-238 in its atmosphere and measure the amount of this and the various decay elements, using spectroscopy, the most powerful tool in astronomy.* This is what is horrendously difficult, but it has actually been done for a few stars, including a red giant called HS 1523-0901 which lies about 7,500 light years from Earth in the direction of the constellation Libra. It has an age of 13.2 billion years, plus or minus 3 billion years, reported by Anna Frebel in 2007.

In spite of the remaining uncertainties in the astrophysical estimates, everything matches the cosmological age determination.

....

* And described properly in *Seven Pillars of Science*.

This is an even more profound fact than you may at first realise. The cosmological age of the Universe is determined from large-scale physics (mostly the general theory of relativity). The astrophysical age of the Universe is determined from relatively small-scale physics of how stars work, with no reference to the general theory at all. Yet they give the same answer. Clearly, science works! Physicists had entitled themselves to a pat on the back. But they cannot rest on their laurels, because at the beginning of the 2020s there appears to be a fly in the ointment. There is a small but possibly significant difference between the determinations of the Hubble constant made using the traditional techniques and those made by studying the background radiation, which looks more worrying as the 'error bars' on the traditional measurement have been reduced. Proponents of each measurement technique insist that their numbers are accurate. But everything does not, after all, fit together quite perfectly. Cosmologists refer to this as 'tension' between the two camps, preferring not to use the word 'disagreement'. But this is the kind of fly physicists ought to welcome, since it points the way to new discoveries. And it just may be relevant to the next story I have to tell.

The Expansion of the Universe is Speeding Up

What was there before the Big Bang? How did the Universe begin? How will it end? Improbably, we have at least partial answers to all these questions. And they are all linked to one crucial property of the Universe – its density.

The equations of the general theory of relativity can be run backwards or forwards, to tell us both about the birth of the Universe and its fate. Looking back, they imply that there was a beginning to time and space, when everything that we can detect in the expanding Universe emerged from a point of zero volume and infinite density, a singularity at 'time zero'. Physicists do not believe that this really happened, because quantum effects do not allow such things to exist. But they do accept that something happened to produce a region of enormous density in a tiny volume that developed into the Big Bang. If we accept that the Big Bang was the time when the entire Universe was at the density of an atomic nucleus, which is the usual rule of thumb, it happened about one ten-thousandth of a second after the beginning of time; half an hour after time zero the temperature was still 300 million K, twenty times the temperature at the heart of the Sun. Everything that has happened to matter since the first one ten-thousandth of a second is explained by well-understood physics. But that first split second is crucial to an understanding of what happened later. And it is intimately related to a profound puzzle about the density of the Universe today.

Although the Universe is expanding, gravity is trying to slow the expansion and pull everything back together. Until about twenty years ago, this seemed to point to a few simple possibilities. Whether gravity will succeed in overcoming the expansion depends on how fast the Universe is expanding and how much matter it contains – its density. If the density is low enough, gravity will be too weak to stop the expansion, and it will carry on for ever. If the density is high enough, gravity will win, and the Universe will eventually stop expanding and begin to collapse back towards a singularity. And there is a unique special case, the so-called critical density, where the expansion gets slower and slower but never quite stops. But it turned out that this is not the whole story.

As Einstein explained, there is a relationship between mass and the way space and time are curved. In the language of the general theory, the ever-expanding version is said to be open, the recollapsing version is said to be closed, and the critical version is said to be flat. The curious thing about our Universe is that as far as we can tell it is, indeed, flat. Why is this curious? Because of what it implies about conditions at the time of the Big Bang, and because absolute flatness is the least likely of all the possibilities, requiring fine-tuning on a mind-boggling scale.

The density is defined by a number called the density parameter, with a value of 1 corresponding to the single critical density, smaller numbers to the infinite number of open possibilities and higher numbers to the infinite number of closed possibilities. Even before the advent of satellites that studied the background radiation, simply by counting the number of galaxies seen in the volume of space we can observe, astronomers knew that the density parameter today must have a value between 0.1 and 10. This sounds like a big range. But the parameter has changed as the Universe expanded away from the Big Bang, because of the changing balance between the density and the expansion rate, which each got smaller but at different rates. As a result, the way the Universe has expanded since the Big Bang continually pulled it away from the critical density. In order for that to lie between 0.1 and 10 today, it had to be precisely 1 to an accuracy of one part in 10^{60} at the time of the Big Bang. The satellite data made things even more puzzling, by revealing that the flatness parameter is indistinguishably close to 1 even today, so it must always have been indistinguishably close to 1 – or, not quite always. 'Only' since the first ten-thousandth of a second. And in that tiny fraction of a second lies the explanation for this improbable feature of the Universe.

As I have hinted, the effects of quantum physics become important for the entire Universe in the interval between time zero and the Big Bang. Among other things, this tells us that there is a quantum of time, the smallest amount of time it is possible to have. This is 10^{-43} seconds. The expansion of the Universe started not from a singularity at time zero ($t = 0$) but at that time, $t = 10^{-43}$ seconds, from a seed no bigger across than the so-called Planck length (10^{-35} m), when the density was not infinite but 'only' some 10^{94} grams per cubic centimetre.* These are the absolute limits on size and density allowed by quantum physics. You might think that such a tiny object with such a huge density would be crushed by gravity and disappear. But at the end of 1979 an American physicist, Alan Guth, realised that need not happen, and found a way to bridge the gap between the beginning of time and the Big Bang.

He noticed that a quantum process called symmetry breaking, which occurs under such extreme conditions, could release energy in the first split-second of time, providing a violent outward push which expanded the Universe so rapidly that gravity did not have time to make it collapse. The violent push soon turned off, but it produced the Big Bang and left the Universe to continue coasting outwards with gravity now able to start slowing the expansion. The release of energy involved in symmetry breaking is like the latent heat released by water when it changes from a vapour to a liquid, but much more extreme. At the beginning of time this process took each tiny region of space, far smaller than a proton, and 'inflated' it to the size of a basketball. And this made the Universe flat.

You can see how this happened from another analogy. The amount of inflation that happened just before the Big Bang was equivalent to taking a tennis ball and inflating it to the size of the entire visible Universe. A tennis ball is obviously curved, bent round into a sphere. But if it were as big as the Universe we see around us, any creatures moving around on its surface would think it was

---

* The seed itself was probably a so-called quantum fluctuation, popping into existence out of nothing at all. See https://www.amazon.co.uk/Before-Big-Bang-Kindle-Single-ebook/dp/B00T6L43NY

flat. In the same way, cosmic inflation has made the space of our Universe indistinguishable from flat space, with the density of the Universe indistinguishable from the critical density.

There's a bonus. In the late stages of inflation, quantum fluctuations in what Alan Guth has called 'the prequel to the Big Bang' produced tiny irregularities. These irregularities grew with the inflation, and were left as irregularities in the Big Bang, acting as the roots from which galaxies and clusters of galaxies could grow. This kind of fluctuation produces a characteristic pattern of bigger and smaller irregularities – and the satellites have shown exactly this kind of pattern imprinted on the background radiation. Even COBE detected this pattern, as well as showing that the Universe is flat, confirming the predictions of inflation theory. The Universe has precisely the critical density of matter. But that posed a puzzle in the 1990s. Where was all the matter needed to flatten the Universe?

By the 1990s, astronomers were well aware that as well as all the bright stars we can see in the Milky Way and other galaxies there must be other stuff that we can't see, dark material which does not shine by its own light but reveals its invisible presence by its gravitational influence on the way galaxies rotate and move around within clusters. But it had taken them a long time to accept the evidence for this.

The speed with which galaxies are moving relative to one another in clusters is worked out from another version of redshift, this time a genuine Doppler effect caused by motion through space. The first person to draw attention to this was the America-based Swiss astronomer Fritz Zwicky, in the 1930s. The random speeds of

Vera Rubin
Getty Images

individual galaxies in a cluster may be more than 1,000 km per second, and they are only prevented from escaping from the cluster by the gravitational pull of all the matter inside the cluster. There has to be more than a certain amount of matter, otherwise the escape velocity from the cluster would be less than the speeds of the galaxies, and the cluster would evaporate as galaxies escaped. When Zwicky tried to balance the equations he found that bright galaxies provide just a small fraction of the mass of a typical cluster. Most astronomers found this so improbable that for decades they simply ignored Zwicky's findings.

Things started to change in the 1980s, after the American astronomer Vera Rubin and her colleagues studied the way individual galaxies rotate, by measuring the Doppler speeds of stars and other features at different distances out from the centre of each galaxy. They expected to see them rotate in the same way the Solar System does, with objects closer to the centre moving faster than ones toward the edge. This happens in the Solar System because most of the mass is concentrated in the centre, in the Sun, so planets further out feel a weaker pull. Because most of the stars, dust, and gas of a galaxy is concentrated in the middle, it seemed obvious that the stuff on the edges shouldn't feel much pull and should move more slowly than stuff near the centre. But Rubin found that disc-shaped galaxies like our Milky Way rotate at the same speed all the way out to the edge of the visible bright disc of stars. The only possible explanation was that the galaxies were each being held in the grip of huge halos of 'dark matter', containing about ten times as much mass as in the bright stars. Zwicky was vindicated. But (there always seems to be a but) this couldn't be the end of the story.

You might think that all this dark matter might be in the form of gas and dust, made up from atoms and molecules, just like ourselves, the Solar System and the bright stars like the Sun – so-called baryonic matter, all built up from protons, neutrons, and electrons. Some of the mass needed to stop galaxies escaping from clusters is indeed in the form of hot gas, which emits X-rays that are detected by satellites. But that isn't the whole story. The well-understood physics of the Big Bang, when conditions were 'only' as extreme as inside the nuclei of atoms today, sets a strict limit on the number of baryons that could have been involved in interactions during the Big Bang. This tells us that the density of baryons in the Universe is no more than 5 per cent of the critical density needed to make the Universe flat. Applying this rule of thumb in the case of clusters of galaxies, there is a limit to how much baryonic matter there can be in a cluster. Even the combined mass of gas and galaxies and any other baryonic matter up to the limit allowed by this rule is still much less than the total cluster mass, showing that there is a great deal of other stuff, non-baryonic matter, around. Since this stuff is both cold and dark, it is known, logically enough, as cold dark matter. Nobody knows what this stuff is, but it is referred to as 'CDM' to make it sound more familiar. The important point is that it is not the same as the stuff we are made of, and it does not interact with baryonic matter except through gravity – it is dark because it does not interact with light or other electromagnetic radiation (such as radio waves or X-rays) at all. This is why it is not subject to the same Big Bang limits as baryonic matter, but that makes it very hard to detect directly, which is why we do not yet know what particles of CDM are like.

Here's the snag. By the mid-1990s it was clear that putting everything together, in order to hold clusters of galaxies together about 5 per cent of the critical density could be supplied by baryons, and about 25 per cent by cold dark matter. This only adds up to 30 per cent of the mass density needed to make the Universe flat. Where is the rest? The few cosmologists who worried about this in the 1990s called the problem the baryon catastrophe. But there was a solution. As I wrote in 1996,* summing up work by David White and Andy Fabian, of the Institute of Astronomy in Cambridge, 'if cosmologists wish to preserve the idea of a spatially flat Universe, as predicated by theories of cosmic inflation, they may have to reintroduce the idea of a cosmological constant'. The reason is that a cosmological constant, like the one Einstein introduced and then rejected when the Universe was found to be expanding, corresponds to an energy, or field, that fills space, and gives it a kind of springiness. This can act like a stretched spring, opposing the expansion of space and holding things back (the way Einstein originally thought of it) or like a squashed spring, pushing outwards against the influence of gravity and making the expansion faster, depending on the exact value of the constant. The Greek letter lambda (Λ) is used to denote the cosmological constant, which is sometimes referred to as a 'lambda field'. As mass and energy are equivalent, the lambda field affects the curvature of space. If the Universe is flat, and it contains only about 30 per cent of the matter needed to make it flat in the form of baryons and cold dark matter, then there is room for 70 per cent of the critical density to be in the form of a cosmological constant,

---

* *Companion to the Cosmos*, Weidenfeld & Nicolson, London, 1996.

or 'dark energy' as it became known. In 1996, this was an obscure and improbable suggestion known only to a few cosmologists (and at least one science writer). A couple of years later, the lambda field was all the rage.

It happened as a result of a serendipitous discovery. In the late 1990s two teams of researchers were trying to improve measurements of the Hubble constant by pushing the technology to measure the properties of exploding stars known as supernova type 1a, or simply SN1a. These are pretty rare; in a galaxy like the Milky Way there are only a couple every thousand years or so. But with thousands of galaxies to study, they are often detected, because for a brief time each one shines as brightly as the entire galaxy it lives in. This is a boon to cosmologists, because every SN1a peaks at the same brightness when it explodes. The brightness is calibrated by studying supernovas in nearby galaxies, whose distances have been found by other means. So when a supernova of this kind explodes in a galaxy far, far away its apparent brightness (or faintness) tells us exactly how far away that galaxy is. Measuring the redshift of the same galaxy then gives astronomers a value for the Hubble constant.

The aim of the two teams was to use this technique, averaged over very many galaxies, to study faint and distant galaxies. Because light takes a finite time to travel through space, when we look at distant galaxies we are seeing light which left them long ago – billions of years ago – when the Universe was younger and more compact. Because gravity tries to hold back the expansion of the Universe, the researchers expected that by comparing their studies of very distant galaxies harbouring SN1a with similar studies of nearby galaxies they would find out how much faster it was expanding in the past, before gravity had had much time to slow it down. To their utter astonishment, in 1998 both teams found that their measurements implied that the Universe was expanding more *slowly* in the distant past – or, putting it the other way round, that it is now expanding faster than it used to. The expansion of the Universe is accelerating.

It is just as well that two teams independently came up with the same conclusion, using entirely independent observations, because to most astronomers this seemed so improbable that if only one team had announced the discovery it would probably have been regarded as a mistake. As it was, the mysterious acceleration was attributed to the phenomenon of dark energy (like CDM, simply a term to disguise the fact that nobody knew what it was), and theorists briefly had a field day trying to come up with exotic explanations for it.* But there was no need. The simplest and most likely explanation for dark energy was already under their noses. I mean the lambda field, of course; the modern manifestation of Einstein's cosmological constant. As we have seen, this contributes a springiness to the Universe, pushing outwards while gravity pulls inwards. In effect, the baryon catastrophe predicted the accelerating expansion, because the amount of dark energy needed to explain the SN1a results is exactly the same as the amount needed to make the Universe flat. And the way the lambda field works also explains why the Universal expansion has only recently (in cosmological terms) started to speed up.

Apart from its actual value, the key property of the lambda field is that it is not only the same everywhere, but the same at all times.

---

* Some of them still do. Well, it keeps them occupied.

Because it is a property of space, each cubic centimetre of space – not just 'empty space' out there among the stars, but the 'space' occupied by the Sun, the Earth and other material things including yourself – contains the same amount of dark energy even if the Universe gets bigger and there are more cubic centimetres of space. So the outward push provided by the field stays the same as the Universe expands. But the inward tug of gravity weakens as the Universe expands and galaxies get further apart. Just after the Big Bang, the influence of gravity was strong enough to overcome the lambda field and slow the expansion. But there came a time when there was a kind of crossover, as the influence of gravity became weaker than the lambda field. At that time, the expansion started to accelerate. This happened about 4 billion years ago, round about the time the Sun and Solar System formed (but that is entirely coincidental).

It is straightforward to work out how much everyday matter you would need, spread out evenly across the Universe, to make it flat. It is roughly 10^{-29} grams per cubic centimetre, equivalent to just five atoms of hydrogen in every cubic metre of space. Of course, everyday matter is not distributed in this way, it is clumped together in galaxies and clusters of galaxies. But everyday matter only contributes 5 per cent of the required density anyway. More than two-thirds of the critical density comes from the lambda field (aka dark energy), which really does contribute the equivalent of nearly 10^{-29} grams per cubic centimetre evenly across the Universe. There is no way to measure this in laboratories here on Earth, and even a sphere as big across as the Solar System out to Neptune contains only the same amount of dark energy as the amount of energy released by the Sun in three hours.

If this is all there is to the Universe, and nothing changes, then the acceleration will get faster and faster, eventually tearing all material objects apart in what has been called the 'Big Rip'. It is just possible (excitingly so) that the 'tension' referred to in Improbability Two is telling us that what is now the standard model of the Universe, referred to as ΛCDM,* is missing something, and that there is a different fate in store. But I have no intention of speculating further in that direction, and will leave you with a summary of what is now the ΛCDM 'best buy' model of the Universe. The present value of the Hubble constant is 67.4 ± 0.5, the total density of baryons plus CDM is 31 per cent of the critical density, and the lambda field contributes the rest, 69 per cent. About one sixth of the matter density is provided by baryons; the stuff we are made of and everything we can see, feel and touch, as well as everything directly observed with our telescopes, makes up, in round numbers, 5 per cent (one twentieth) of the Universe. If that doesn't sound improbable today, it certainly did 25 years ago when the baryon catastrophe first caught the attention of cosmologists.

---

* Notice that this shorthand term for the standard model doesn't even mention baryons; the stuff we are made of is such an insignificant fraction of the Universe it got left out.

IMPROBABILITY

We Can Detect Ripples in Space Made by Colliding Black Holes

easuring the age of the Universe as 13.8 billion years with an error of no more than a hundred million years – less than 1 per cent – is, to say the least, impressive. But at the other end of the size scale, physicists can measure displacements in detectors 4 km long that amount to roughly one ten-thousandth of the width of a proton. This seemingly impossible (but actually only highly improbable) achievement was necessary for them to be able to detect ripples in space predicted by Einstein's general theory of relativity – gravitational waves.

A useful way to think about the relationship between matter, space, and gravitational waves involves imagining a heavy weight placed on a stretched rubber sheet like a trampoline and jiggled about. My version of the story has appeared with minor variations in several of my books,* so look away now if you don't want to see it

---

* Most recently here: https://www.amazon.co.uk/Discovering-Gravitational-Waves -Kindle-Single-ebook/dp/B071FFJT74

again. The key thing to think about here is how fast a gravitational influence reaches out across the Universe.

The presence of any object that has mass distorts space around it, and we can represent a mass like the Sun as a bowling ball plonked on the hypothetical trampoline. The ball makes a dent on the surface, and marbles rolled across the surface follow curved lines around the dent. Similarly, the curved space around a massive object such as the Sun makes things (even light) follow curved paths, as if there was a force (gravity) tugging them towards the Sun. It was Einstein's prediction of how much starlight would be bent as it passed near the Sun that enabled astronomers to confirm the accuracy of his general theory during a solar eclipse in 1919, making Einstein famous.

But what if the bowling ball is taken away? The curved surface of the trampoline becomes flat again, but it doesn't do so instantly. The smoothing out spreads across the surface. The Earth is following an orbit around the Sun because of the dent the Sun makes in spacetime.* If the Sun suddenly ceased to exist, the Earth would not immediately fly off into space, because the dent would still be there for a while, until there had been time for the news that the Sun had gone to reach us. Einstein already knew, from the special theory of relativity, that nothing could travel faster than light; so he expected that gravity would travel at the same speed. If the Sun disappeared, the Earth would continue in its orbit, and the sky would be bright for another eight and a bit minutes, then the sky would go dark and the planet would fly free at the same time.

....

* I have to be careful here not to say that it is just a dent in space; the situation is a bit more complicated than the bowling ball analogy, but I shall not go into the details.

But remember the bowling ball being taken off a trampoline. The stretched surface doesn't instantly go back to being flat; it bounces up and down for a bit as it settles down, sending ripples across the surface. If the Sun disappeared, the space (spacetime) around it would presumably ripple in the same way, with the ripples dying down while it smoothed out. The ripples would be gravitational waves. After a false start when he made a mathematical slip, Einstein published the idea in 1918. But he was never quite sure if the effect was real, and once said: 'If you ask me whether there are gravitational waves or not, I must answer that I do not know. But it is a highly interesting problem.' Exactly a hundred years after he published the idea, however, such ripples were detected on Earth for the first time.

The detection required a huge effort, but one reason why physicists were sure the effort was worthwhile was because by then they had direct evidence of the effects of gravitational radiation on the behaviour of pairs of stars known as binary pulsars. Pulsars are rapidly spinning neutron stars – balls of matter only about 10 km across but with the density of an atomic nucleus, containing about as much mass as our Sun, left over when some stars much bigger than the Sun explode as supernovas. We can detect them because they have strong magnetic fields and beam out radio waves, like the beam from a lighthouse. Some of these beams flick across the Earth and can be detected, but there must be many pulsars whose beams are not oriented in the right direction for us to see them.

In 1974 Russell Hulse, a PhD student at Harvard University, was using a giant radio telescope at Arecibo in Puerto Rico (it featured in the movie *Contact*) to carry out a search for pulsars, under the supervision of Joseph Taylor. On 2 July 1974 he found a pulsar right at the limit of the telescope's capacity to make identifications, and after checking carefully over the next few weeks, he confirmed that it was a genuine discovery, and labelled the object PSR 1913+16. It proved to be an extreme example of its kind. The neutron star was spinning once every 58.98 milliseconds, making it the second-fastest pulsar known at the time, so that the beam produced seventeen blips in Hulse's detector every second.

But as Hulse continued to observe the pulsar, he found that it was changing in what seemed to be an impossible way. The measurements revealed a complicated pattern of behaviour. Sometimes the pulses arrived a little sooner than expected; sometimes a little later than expected. These variations changed smoothly and over a repeating period of 7.75 hours. Hulse realised that the changes could only be caused by the pulsar orbiting around another star. The speed with which the changes were taking place showed that the orbit of PSR 1913+16 must be tiny, which meant that its companion must be tiny – another neutron star. The two stars must actually form a binary pair, with similar masses, each star orbiting around their common centre of mass. So it became known as 'the binary pulsar', although only one of the neutron stars is detected as a pulsar.

This is an extreme system, making it an ideal testbed for the predictions of the general theory. Continuing observations showed that as the pulsar orbits its companion once every 7 hours 45 minutes, with an average speed of 200 kilometres per second, it reaches a maximum speed of 300 kilometres per second – a thousandth of the speed of light. The distance round the orbit is roughly 6 million kilometres, which is, coincidentally, about the same as the circumference of the Sun. So if the orbit of the binary pulsar were circular the whole system would fit inside the Sun, with the two neutron stars about the same distance apart as the distance from the centre of the Sun to its surface. As it happens, though, the orbits are elliptical, so the *pas de deux* danced by the two stars is more complicated. At their closest the two objects are about 1.1 solar radii apart; at maximum separation they are 4.8 solar radii apart. This is almost a textbook setup for producing gravitational waves.

You can see why if you imagine two hollow metal spheres, connected by a short rod, floating in a tank of water. If they do not move, there will be no waves in the water. But if they are rotating around one another, like a spinning dumbbell, waves will ripple out across the surface. The same sort of thing happens to space-time when two neutron stars separated by less than the diameter of the Sun are orbiting around one another. But making waves requires energy. As energy from the binary pulsar is going into the gravitational waves, the two stars have to spiral together to give up gravitational energy, moving faster as they do so. The orbital period will decrease ('decay') by a tiny amount which can be precisely calculated from the general theory.

The prediction was that the orbital period of the binary pulsar, which is about 27,000 seconds, would decrease by about 0.0000003 per cent, or 75 millionths of a second, each year. In order to measure such a tiny effect, the astronomers had to make allowances for all kinds of influences, including the motion of the Earth in its orbit around the Sun and changes in the rotation of the Earth itself. Taking all these effects into account, after analysing roughly 5 million pulses from PSR 1913+16, in December 1978

Taylor was able to announce that the orbit of the binary pulsar was decaying exactly in line with the predictions of the general theory. The general theory is right, and gravitational waves are real. More than 50 binary pulsars are now known, providing even more evidence in support of the accuracy of the general theory, but the one discovered by Hulse and Taylor is still referred to as 'the' binary pulsar.

By the end of the 1970s, there was no doubt that gravitational waves do exist. But that left the enormous challenge of detecting gravitational waves directly, here on Earth. To most people, it looked impossible, because by the time waves from something like a binary pulsar reach us the ripples are much, much smaller than the size of an atom. But there are cosmic events which, theory predicts, should produce significantly bigger waves, and that offered a chink of hope for the experimenters.

That hope rested upon the possibility of detecting gravitational waves using the technique of interferometry. This literally depends on the way two things (such as light beams, in the laboratory version of such experiments) interfere with one another. Here comes another of my familiar analogies. When a pebble is tossed into a calm pond, ripples spread out smoothly in all directions. But if you toss two pebbles into the pond at the same time, you get two sets of ripples which interfere with one another, making a more complicated pattern. In some places the waves cancel out to leave the surface more or less flat; in other places the waves add together to make extra high ripples. This process may be familiar to you from a classroom experiment with light, designed to demonstrate how it behaves like a wave. In a darkened room a beam of light is shone through two small holes in a screen (a piece of paper or card is good enough for the job) and shone on to a second screen. The light waves spreading out from each hole in the first screen interfere just like those ripples on a pond, producing a pattern of light and shade on the second screen – an interference pattern. Physicists realised that in principle this kind of interference could be used to measure very small changes that would be produced by a gravitational wave squeezing and stretching the space between two objects. But putting the principle into practice would be difficult and expensive. The long saga involving a mixture of politics, science and personality clashes that followed this realisation has been entertainingly described by Janna Levin, in her book *Black Hole Blues*;* but I shall cut to the chase and describe the outcome of all these shenanigans.

Interferometry can be applied to the search for gravitational waves because of the way these waves distort spacetime. They do not produce ripples in the direction the wave is moving, the way water waves do, but change the shape of space at right angles to the direction the wave is moving. This squeezes space in and out in a regular way. When one direction is being squeezed the direction at right angles to the squeeze is stretched, and vice versa. So physicists realised that if they had a detector with two arms at right angles to each other, like a capital letter L but with both arms the same length, a gravitational wave passing through it would squeeze one arm at the same time it was stretching the other. These changing lengths provide a characteristic 'signature' of gravitational

* Bodley Head, London, 2016.

waves, which could be monitored using interferometry, if the arms were long enough and the detectors sensitive enough.

The light required for the job has to come from lasers, which produce very pure beams with very precise wavelengths. The laser light has to be split into two beams, which are precisely in step with one another; these are then sent along the two arms of the experiment at right angles to each other but precisely the same length, before being reflected back along the same paths to merge again and make an interference pattern, monitored by an automatic system. If the experiment is perfectly set up, the returning waves will cancel each other out, and the monitoring system will detect nothing at all. But when a gravitational wave passes through the experiment, one arm is squeezed and the other stretched, so that the two beams get out of step. The resulting interference can be recorded, and even displayed on a monitor screen as wiggly lines, equivalent to the pattern of light and shade produced in the classroom experiment with two holes.

An obvious question is, how do the laser beams detect the stretching and squeezing of space, when they also are affected by the gravitational waves, being stretched and squeezed just like everything else? The answer lies in the fact that we are really dealing with spacetime, not just space. The distortion of spacetime affects how long it takes for the light beams to get from one end of the experiment to the other. What the interferometer actually measures is a time difference, not a space difference; but it is a simple matter to convert that into the spatial equivalent.

A formal proposal for a gravitational wave detector was put forward in the USA in 1983. Nothing short of ambitious, it proposed a pair of identical detectors at widely separated locations. The idea was that gravitational waves would affect both detectors in the same way, with a small time delay, which meant they could be distinguished from local disturbances which affected each individual detector. The original proposal envisaged that each detector would have arms 10 km long, and the project would cost $70 million. The American National Science Foundation gave it the go-ahead in 1986, but the size of the arms of the two detectors had to be limited to 4 km, because the available sites were not big enough for anything more. Construction began in the mid-1990s, but to nobody's surprise, although the detectors were smaller the costs got bigger, soaring above $1 billion. Arguably, the most improbable feature of the whole project is that it ever got funded! The detectors were built at sites just about as far apart as possible within the contiguous United States, at Hanford, in Washington State, and Livingston, in Louisiana. What became known as the Laser Interferometer Gravitational-wave Observatory (LIGO) was the most expensive project ever funded by the NSF, which meant that there were many sighs of relief when, improbably, it actually discovered something, in September 2015. But the way in which the discovery was made is as intriguing as the fact of the discovery itself.

While the detectors were being built and tested, the theorists worked out exactly what kind of 'signal' they might detect, using computer simulations based on the general theory of relativity.* Their best bet was the collision and merger of two black holes. This

....

* Astronomers use the term 'signal' to refer to any burst of light or other radiation from space, such as the radio noise from pulsars; there is no implication that an intelligence is producing it.

The LIGO detection site at Livingston, Louisiana
Caltech/MIT/LIGO

occurs inevitably in any binary system similar to the binary pulsar (indeed, it will inevitably happen, eventually, to the binary pulsar itself) as its component parts spiral together, but black holes are more massive than pulsars and will produce a bigger burst of waves. But while they were at it, the theorists also calculated what might be observed when two neutron stars merge. Simulations of how a pair of black holes would spiral together and merge were carried out for a variety of different black hole masses. They discovered that, if Einstein's equations are right, such a merger will produce a distinctive feature which they called a 'chirp', of gravitational waves, in which the ripples get shorter and shorter in wavelength (higher in pitch, in musical terms) as the black holes get closer and closer together, then ending abruptly as they merge into one object. In audible terms, this would be like the sound you make by running your hand rapidly along the keys of a piano from left to right. So the experimenters knew exactly what they were looking for. But as if their work wasn't hard enough already, they had set themselves a deadline. Einstein had finished his general theory in November 1915 and it was formally published early in 1916, which was also the year he first found a hint that the theory predicts the existence of gravitational waves. The LIGO team decided that it would be rather nice if they could detect such waves in 2016, a hundred years after the publication of the general theory. Improbably, and to their own surprise, they did better than that.

The details of the detector systems are mind-boggling. Light from a 20-Watt laser goes down each of the 4-kilometre-long arms through an evacuated tube a metre in diameter. At the ends of the tubes special partially reflecting mirrors bounce the light to and fro about 280 times before it is released into the interferometer set-up, effectively boosting the power of the detector. But since the prediction of the theorists was that the waves they were looking for would change the mirror spacing by only about 10^{-18} m, less than one thousandth of the diameter of a proton, the mirrors had to be shielded from any form of outside vibration, ranging from traffic on nearby roads (including staff bicycling in to work) to the movement of weather systems on the other side of the continent, the movement of currents in the Pacific Ocean, and just about every significant earthquake on Earth.

This was achieved by suspending each of LIGO's 40 kg test masses (heavy weights to which the mirrors are attached) in a system of four pendulums. Part of this suspension was 'passive', simply allowing the building to move around it while the test masses hung below. But the really clever bit was an 'active' system, which measured seismic disturbances and very gently pushed the other way to cancel them out, the way noise-cancelling headphones respond to sounds from outside and cancel them out.

With everything in place and tested, the first proper science run of the detectors was planned for September 2015. In order to be ready, a test run to make sure all these systems were working was being carried out in the middle of the night on Monday, 14 September. During a pause in those tests, the detectors were left in observing mode, although nobody was expecting to observe anything. But at 2.50am local time in Hanford, and 4.50am local time in Livingston, almost simultaneously, each detector recorded a chirp lasting 200 milliseconds. The detectors had picked up a gravitational wave signal far stronger than anyone had expected, and far more quickly than they had anticipated. Because there was a delay of just 6.9 milliseconds between its arrival at the first detector and its arrival at the second detector, this confirmed that the wave travelled at the speed of light.

The details of the chirp match the predictions for the spiralling together and merger of two black holes, one with about 29 times the mass of our Sun and the other about 36 times the mass of the Sun, to make a single black hole with a mass of about 62 times the mass of the Sun. The 'lost' mass tells us that about three times the mass of our Sun was converted into energy in the form of gravitational waves in the process. This is equivalent to 10^{23} times (a hundred billion trillion times) the luminosity of the Sun.

After checking and double-checking their observations to make sure there was no mistake, the team officially announced their discovery on 11 February 2016, almost exactly a hundred years after the general theory of relativity was announced to the world. But even before the news broke, LIGO detected a second black hole merger, which shook the detectors on Christmas Day 2015.* It was caused by the merger of black holes with fourteen and eight times the mass of the Sun, combining to make a black hole with a mass 21 times that of the Sun, with about one solar mass of matter being converted into energy. This showed that the first detection was not a fluke.

Since 2015, gravitational wave astronomy has become almost routine. The detection of another black hole merger is no longer

---

* Christmas Day in the United States. Astronomers generally quote times in a system which is essentially the same as GMT; that puts it in the early hours of 26 December 2015, so it is often referred to as the Boxing Day event.

news, just as the discovery of another planet orbiting a distant star is no longer news. But one other kind of merger is worth mentioning.

There is now a third gravitational wave observatory operating, a European detector called Virgo (after the constellation), similar to LIGO. With three working observatories on the ground, scientists can more precisely identify the region on the sky where gravitational waves come from. Similar detectors will soon come online in Japan and India, but three was enough to make a spectacular discovery in the summer of 2017. In August that year, all three gravitational-wave detectors saw a signal identified as a binary neutron star merger that occurred between 85 million and 160 million light years away. The combined mass of the two colliding stars was about three times the mass of our Sun. Because they could triangulate the source, the gravitational wave astronomers were able to tell other astronomers where to point their telescopes to see if they could spot anything interesting associated with the event. Within hours, five groups had identified a new source of light in a galaxy known as NGC 4993. This faded from bright blue to dim red over the next few days, and a couple of weeks later it began to emit X-rays and radio waves. Spectroscopic studies of the fading light showed that the violent outburst associated with the neutron star merger (called a hypernova) had produced huge quantities of heavy elements, including, to the delight of headline writers, gold. This solved a long-standing mystery.

As I described in *Seven Pillars of Science*, before 2017 astronomers knew that gold and other heavy elements could be produced in another kind of stellar explosion, supernovas; but they also knew that it was impossible for these events to make all the heavy elements we see in the Universe. Hypernovas turn out to be able to make just enough heavy elements to fill the gap. The event seen in August 2017 alone produced between three and thirteen Earth masses of gold, and similar events account for at least half the gold that there is in the Universe today. Improbable though it may seem, this means that much of the gold in any jewellery you possess was manufactured when neutron stars collided and merged.

IMPROBABILITY

 *

Newton, the Bishop,
the Bucket, and the Universe

This seems like a good place to stand back from cutting-edge twenty-first-century science and catch our metaphorical breath by looking at a more philosophical (or metaphysical) improbability, which taxed the brain of no less a thinker than Isaac Newton. As well as Newton, it involves a bucket of water, a long rope, and a bishop – which sounds like the beginning of a joke, but actually leads to a Deep Truth about the nature of the Universe, which helped Albert Einstein to develop the general theory of relativity.

You don't actually need a bucket to get a handle on what the puzzle is all about. It is visible in all its glory every time you stir cream into your coffee and watch the pretty patterns it makes as it swirls around. How does the cream, and the coffee, 'know' it is swirling

---

* This section revisits some ideas discussed in the first edition of my book *In Search of the Big Bang*, but left out of later editions as they were regarded as too much of a diversion from the main story there. Suitably revised, though, they seem to fit perfectly here!

around? You might guess that it is because it is moving past the side of the cup. But there is far more to it than that.

Newton was the first person to realise the implications (and for all I know he did get the idea from watching coffee swirling in a cup, since coffee drinking was very fashionable in his day). But Galileo seems to have been the first person to point out a closely related feature of the world – that it is not the speed with which an object moves but its acceleration that reveals the presence of forces acting upon it. On Earth, because of friction, wind resistance, and other influences that can never be got rid of, there are always outside forces trying to slow down a moving object. We have to keep pushing to keep it moving. But in space, as we have all seen in TV broadcasts by astronauts, things keep moving in a straight line until they feel the effect of a force. The nearest we can come to this on Earth is the motion of an ice hockey puck skimming across the ice, or (slightly less like motion in space) the puck on an air hockey arcade game. This really does seem to keep moving in a straight line at constant speed (that is, at constant velocity) until an outside force interferes with it. It was Newton who, without the benefit of ever seeing those TV broadcasts, worked out the law that the acceleration produced by a force is equal to the force divided by the mass of the object, and extended this (with the aid of his law of gravity) to explain the orbits of the planets around the Sun. In modern terminology, a frame of reference in which things move with constant velocity unless acted upon by external forces is called an inertial frame, and Newton had the idea that there must be some fundamental inertial frame, an absolute standard of rest, which is somehow determined by empty space. He argued that things move at constant velocity *relative to empty space* unless they are accelerated by outside forces.

There's an obvious snag with this idea. How do you know what empty space is? You can't hammer a nail into it and then measure all your velocities relative to the nail. How can you identify this absolute standard of rest? This is where the bucket comes in. Newton thought that he could use it to identify the fundamental inertial frame, as he described in his great book the *Principia*:*

The effects which distinguish absolute motion from relative motion are, the forces of receding from the axis of circular motion ... if a vessel, hung by a long cord, is so often turned about that the cord is strongly twisted, then filled with water, and held at rest together with the water [then let go]; thereupon, by the sudden action of another force, it is whirled about the contrary way, and while the cord is untwisting itself ... the surface of the water will at first be plain, as before the vessel began to move; but after that, the vessel, by gradually communicating its motion to the water, will make it begin sensibly to revolve, and recede by little and little from the middle, and ascend to the sides of the vessel, forming itself into a concave figure (as I have experienced[†]), and the swifter the motion becomes, the higher the water will rise.

....

* The translation of Newton and the quotation from Berkeley below are taken from *The Unity of the Universe*, Dennis Sciama, Faber & Faber, London, 1959.
† An important point. Newton actually did the experiment, it wasn't just an imaginary 'thought experiment'.

If the 'vessel' is then grabbed suddenly and held still, the water inside will still be rotating, and still rise up the sides, only gradually slowing down and flattening out, like the coffee stirred in your cup. It is not the relative motion of the bucket and the liquid that matters, but the absolute motion of the liquid relative (Newton thought) to empty space. When the bucket is rotating but the water is still, the surface is flat. When the water is rotating and the bucket is still, the water surface is curved. And when both are rotating, with no motion relative to one another, the surface is curved. In a modern version of the experiment, you could stand your cup of coffee dead centre on a turntable and watch it rotate. Both cup and liquid would be rotating, but the liquid would still form a concave surface. The liquid 'knows' it is rotating and acts accordingly. Philosophers who objected to Newton's argument and to the idea of absolute space on the grounds that something completely unobservable cannot be real had to find another way to explain what was going on in Newton's bucket. It took them 30 years, but then George Berkeley, an Irishman who was born in 1685 (the year before the publication of the *Principia*) and grew up to become a philosopher, economist, mathematician, physicist, and bishop (not necessarily in that order, to paraphrase Eric Morecambe) came up with an answer.*

Berkeley said that all motion must be measured relative to something. Newton's 'absolute space' is literally nothing, and cannot be perceived, so it doesn't fit the bill. As he put it, 'there is no space where there is no matter'. He pointed out that if everything in the Universe were annihilated except for a single globe (let's call it

---

* Berkeley, California, was named in his honour.

the Earth), then it would be impossible to envisage any motion of that globe, either through space, or as a rotation. There would be nothing to measure against, so movement would be meaningless. Even if there were two perfectly smooth globes in orbit around one another, there would be no way to measure that motion. But 'suppose that the heaven of fixed stars were suddenly created and we shall be in a position to imagine the motions of the globes by their relative position to the different parts of the Universe'. So 'it suffices to replace absolute space by a relative space determined by the heaven of fixed stars'. According to Bishop Berkeley, the coffee stirred in your cup rises up the sides because it knows that it is rotating *relative to the distant stars*. There's another example which many people have seen today. Science museums often contain a great Foucault pendulum, swinging ponderously to and fro. Such a pendulum keeps swinging in the same plane, while the Earth rotates underneath it, so that to us it seems as if the plane of the pendulum is rotating while we are standing still. The pendulum 'knows' where it is relative to, in Berkeley's terminology, the fixed stars, and holds itself steady compared with them and (we now know) the Universe beyond the stars.

But how does it know? What mysterious influence can reach out across space to affect the motion of water in a bucket, or a swinging pendulum, here on Earth? With no answers to these questions, Berkeley's idea did not gain wide currency during his own lifetime or in the century and a half that followed. But it was revived in the second half of the nineteenth century by Ernst Mach, the physicist whose name is immortalised in the number used to measure speed relative to the speed of sound.

Ernst Mach
Alamy

Mach was born in 1838 in what was then the Austro-Hungarian Empire. He earned his PhD at the University of Vienna, and became a professor at the University of Prague in 1867 before returning to Vienna in 1895. He died in 1916, the year that Einstein published his general theory of relativity; but there is more than this calendrical coincidence to link him with that theory. It was while he was working in Prague that in 1883 Mach published a book, *Die Mechanik*, which took the puzzle of absolute motion a stage further, denying the reality of either 'absolute space' or 'absolute time', and that was a key influence on Einstein when he was developing the general theory. Mach wrote that '[when] we say that a body preserves unchanged its direction and velocity *in space*, our assertion is nothing more or less than an abbreviated reference to *the entire universe*'.*

Before we look at the link between that idea and the general theory, there's another key point to take on board. There are actually two kinds of mass. One kind comes in to Newton's equation relating force, mass, and acceleration. It is a measure of how much an object resists being pushed about, and it is called inertial mass. The other kind of mass determines how strongly an object is tugged by gravity, and how strongly it tugs on other objects. This is called its gravitational mass. The gravitational force of attraction between two objects is actually proportional to their two gravitational masses multiplied together and divided by the square of the distance between them – Newton's inverse-square law of gravity. The intriguing thing is that for any particular object both masses,

* His emphasis.

gravitational and inertial, are the same. You can measure the gravitational mass of an object by measuring how strongly it is tugged by gravity (simply by weighing it as it is tugged by the Earth) *or* measure its inertial mass by pushing it with a known force to see how quickly it accelerates. These are completely independent tests, and they always give the same answer as each other. Which means there must be some profound link between inertia and gravity. It is also something you see the effect of every day – the equivalence of the two masses is the reason why all objects fall at the same rate.

Newton thought that inertia was intrinsic to an object. In an otherwise empty universe, a single sphere would have the same inertia that it has in our Universe. Mach reasoned that inertia is caused by the 'fixed stars' – that is, by the presence of all the other matter in the Universe. Take them away and the lonely globe would have no inertia. This led him to a curious conclusion, involving the equatorial bulge of the rotating Earth.

In everyday language, the equatorial bulge is said to be caused by centrifugal force. All self-respecting physicists hate the term, because there is no such thing. It is a 'fictitious force' caused by circular motion. What is actually happening is that the material at the surface of the Earth near the equator would keep moving in a straight line if it were not being pulled inward by the Earth's gravity. This inward force (a centripetal force) is what stops the planet flying apart, but if someone invented a machine that cancelled out gravity, the machine would not fly outwards like Cavor's spaceship in H.G. Wells' *The First Men in the Moon*, but off on a tangent to the surface of the Earth. If the planet spun fast enough for bits to break off from the equator, that is what would happen to them.

The equatorial bulge is best explained in terms of the energy of this motion and the gravitational energy involved, but as I don't want to get sidetracked into those details and I lack self-respect, I will accept centrifugal force as the term to use here. What matters is that the Earth does bulge at the equator and that this is because it is rotating. Which is where Mach casually tossed a spanner in the works. What matters, he said, is the *relative* rotation. It doesn't matter whether the Earth is rotating and the stars are still, or whether the Earth is still and the stars are rotating around it. Either way, you will get an equatorial bulge. In his words, 'it does not matter if we think of the Earth as turning round on its axis, or at rest while the fixed stars revolve around it'. Einstein seized on this package of ideas, and gave the notion that inertia is a result of the existence of everything in the Universe the name Mach's Principle. But not everyone liked it – it was denounced by both Vladimir Ilyich Ulyanov (aka Lenin) and Bertrand Arthur William Russell.

The principle of equivalence – that gravitational and inertial mass are identical – is one of the cornerstones of the general theory of relativity, and Einstein tried to make Mach's Principle part of the theory. He argued that this identity between gravitational and inertial mass exists because inertial forces are themselves gravitational in origin.

It is easy to make a vague argument in support of this idea (what physicists call, for obvious reasons, a 'hand-waving argument'). Gravity works both ways (actually all ways, which is what is really important). The Earth pulls on me, but I am also pulling on the Earth. The Earth pulls on the Moon, but the Moon is also pulling on the Earth, so both of them are orbiting around their common centre of mass, which lies about 1,700 kilometres below the surface of our planet, not at its centre. If the 'fixed stars' are somehow reaching out with gravitational fingers to influence the motion of an object here on Earth, then there ought to be a corresponding influence from that object which reaches out to the stars. When we try to move something by making it rotate or accelerating it in a straight line, it is moving through the cosmic gravitational web, and disturbing it like a fly struggling in a spider's web. The result should be a disturbance that spreads out through the web and back to the stars (or galaxies, from a more modern perspective), which send back some sort of reaction, like a cosmic handshake, trying to maintain the status quo, resisting the acceleration and producing inertia.

It sounds fine, until you remember that no signals can travel faster than light. If I push the pencil on my desk, it immediately 'knows' that I am pushing it and how much it should resist that push. Signals moving backward and forward through Newtonian empty space will not do the trick. But Einstein's image of spacetime as a flexible four-dimensional fabric in which matter distorts space, and the distortions in space tell material objects how to move, gives us a different perspective, and a way to tackle the puzzle using the equations of the general theory instead of just waving our hands about in a vague way.

When he was searching for a theory of gravity, from the outset Einstein intended that Mach's Principle should be a natural part of the general theory. When people like me describe the behaviour of things like binary neutron stars in terms of dents in spacetime, we ignore the rest of the Universe and pretend the stars are orbiting one another in an entirely flat background spacetime. But the spacetime at any location is in principle affected by the gravity of every material object in the Universe, because there is no limit to the range of gravity. There's a subtle and often overlooked effect at work here as well. If you put a heavy weight like a bowling ball on a stretched elastic sheet like a trampoline, it makes a single dent. Take the ball away and put another one on a different part of the sheet and it makes a different dent. But if you put both balls on together, the pattern of dips that results is not exactly the same as you would get by adding up the two separate dips, because when you add the second ball it is going on to a sheet that has already been distorted by the presence of the first ball. Imagine the complications for the shape of spacetime caused by adding the effects of every material object in the Universe. The shape of the spacetime in which binary neutron stars – or any objects – move is, in spacetime terms, quite literally not a level playing field.

The equations of the general theory of relativity should include the effects of all those distant masses against which accelerations, inertial forces, and rotations are measured. And they do. But there is a twist. Einstein's equations only produce the right Machian influences in one kind of world, the kind in which the Universe is closed. For decades, this was seen by some as a flaw in Einstein's theory, because astronomers thought that the Universe was open, in the sense described in Improbability Three. But the Universe only has to be *just* closed. It can be as near to flat as you like, provided it is on the closed side of the dividing line. The modern idea of inflation and the evidence from studies of the background radiation that the Universe is indistinguishably close to flatness exactly chime with this requirement of the general theory. Rather than being a flaw, the requirement is actually a triumph!

Even better, there is some experimental evidence that the kind of influence on spacetime predicted by the Machian aspects of the general theory are real. Almost as soon as the general theory was published, a couple of theorists used the equations to work out how local concentrations of matter could, in principle, produce a local equivalent of the Machian influence, called frame dragging. In one variation on the theme, you imagine an object placed inside a large, perfectly smooth spherical shell of matter, and the shell is made to rotate relative to the distant galaxies. If Mach's Principle, as incorporated into the general theory of relativity, is correct, the object inside the shell should feel a small force trying to drag it round with the shell. In another version of the idea, the equations tell us that close to a spinning mass like the Earth there should also be a tiny frame-dragging effect. The calculations involving frame dragging were first carried out in 1918, in the framework of the general theory of relativity, by the Austrian physicists Josef Lense and Hans Thirring; but the effects they predicted were so tiny that nobody expected that the Lense–Thirring effect would ever be measured. But, improbably, it was.

In 2004, a satellite known as Gravity Probe B was launched into orbit around the Earth, carrying four gyroscopes in the form of spheres roughly the size of table-tennis balls, each perfectly round to within less than 10 nanometres, meaning there were no irregularities bigger than 40 atoms in height. By monitoring the spin of these gyros the experimenters, from Stanford University, measured a frame-dragging effect of 37.2 ± 7.2 milli-arc-seconds per year, compared with a prediction from the general theory of 39.2 milli-arc-seconds per year. This triumph was only one of Gravity Probe B's achievements, but that's a story for another book. What matters here is that the prediction has proved correct, suggesting that Mach's Principle is right. Before I leave the story of Newton, the Bishop, the Bucket, and the Universe, though, I'd like to mention a prediction that never was made, but could have hinted that Mach's Principle was correct a hundred years ago.

I have been dancing round the terms 'fixed stars' and 'distant galaxies' because until the 1920s what we now think of as the Milky Way galaxy, an island in space containing hundreds of billions of stars, was thought to be the entire Universe. It then emerged that we live in a flattened disc of stars and that there are other, similar islands (some disc-shaped, some not) beyond the Milky Way. Eventually (but not until right at the end of the twentieth century*) it was established that the Milky Way galaxy is almost exactly average-sized, as disc galaxies go, making it a typical member of its class. We live in an ordinary part of the Universe. By then, it was also well established that there are hundreds of billions of other galaxies in the Universe. Nobody could have predicted that. Or could they?

Although earlier astronomers had studied the band of light across the sky that we call the Milky Way and inferred that we live in a star system shaped like a mill wheel or grindstone, it was only at the end of the second decade of the twentieth century that the American Harlow Shapley was able to put this on a proper scientific footing

------

* See *The Observatory*, Vol. 118, pp. 201–08 (1998).

using what was then the most powerful telescope in the world, the 100-inch reflector on Mount Wilson in California. It would take even bigger and better telescopes to identify other galaxies. But if Shapley had been a Machian maybe he could have inferred their existence.

By 1920, it was clear that the Milky Way galaxy is a flattened disc made up of a myriad of separate stars. Why is it flat? Not because it is a solid object like a grindstone, but because it is rotating. How does it know that it is rotating? Mach's Principle tells us that it is because there must be, far beyond the Milky Way, some distribution of matter which provides a frame of reference against which the rotation of our galaxy is measured. And the natural guess would be that if our Milky Way is an island of stars, the Universe must be filled with other islands of stars – other galaxies. If anybody had made the connection at that time, the work of Edwin Hubble and others which established the existence of other galaxies and the scale of the Universe might have come as less of a surprise, and been seen as confirmation of the accuracy of both Mach's Principle and the general theory of relativity.

It never happened that way, but it is something to think about next time you stir cream into your coffee. As you do so, the coffee rises a little way up the sides of the cup and dips a little in the middle. The improbable possibility is that it does so because it is feeling the influence of hundreds of millions of galaxies hundreds of millions of light years away.

Simple Laws Make Complicated Things, or Little Things Mean a Lot

The two previous improbabilities dealt with the influence of large things on small things. Hugely energetic events in deep space that produce tiny movements of objects on Earth, and the entire material Universe telling small things how to resist being moved. But tiny things can also have a big influence on the world at large. You may have heard of the 'butterfly effect', a term that is about as misused as the expression 'quantum leap'.* But the improbable truth about the butterfly effect goes much deeper than the popular misconception.

From the time of Isaac Newton until well into the twentieth century, the Universe seemed to be an orderly place obeying simple laws in a deterministic fashion. His famous laws told us how things move when they feel a force, and explained the orbits of the planets

---

* A quantum leap is actually the smallest possible change, made in a completely random way. Not quite what advertisers think they mean when they say that a product represents a 'quantum leap' from last year's model.

and in principle the stars. This led to the analogy of the Universe as like a clockwork mechanism, wound up in the beginning and set inexorably along a predictable path into the future. But almost from the time of Newton himself there was known to be a problem with this image, which was mostly ignored in the hope that it might one day be resolved. It involves gravity and orbits, and it is called the 'three-body problem', although it actually applies to the behaviour of any group of gravitationally interacting objects made up from more than two components.

The problem is that although Newton's laws allow us to calculate with absolute precision the orbits of two objects around one another under the influence of gravity, they do not give exact solutions for problems involving three or more gravitating objects. We can calculate the orbit of the Moon relative to the Earth by ignoring any other objects, and we can calculate the orbit of the Earth around the Sun in the same way, but we cannot calculate the combined behaviour of the Earth–Moon–Sun system,* let alone the rest of the Solar System and the Universe at large. This cannot be done *in principle*, it isn't just that the problem is too hard for us to solve. The relevant equations are said to be non-integrable, or to have no analytical solutions.

We can get round the problem, sometimes, using approximations. In this example, pretend that the Earth is not moving and calculate how the Moon moves in a short time interval, then ignore the Moon and calculate how far the Earth has moved in that interval

---

* The word 'system' is used to refer to any set of interacting objects, such as the Solar System.

under the influence of the Sun and calculate the next step in the Moon's orbit, and so on in a repeating series of steps (iteration). But at each step the Moon is also being influenced by the Sun, and will not be exactly where the previous calculation left it. And what about the Moon's influence on the Sun and Earth? In this example, the mass of the Sun is so much bigger than that of the planets that the approximation works well for determining the orbits of the planets and lesser objects in the Solar System; but if all three objects have about the same mass the problem cannot ever be solved analytically. The key point is that because the equations have no analytical solutions the Universe itself does not 'know' how a three-body system will change as time passes.

This sort of thing would not matter if small errors in the calculation always produced small differences in the final result. But that is not always the case, and that is half of the story of chaos. A system in which small differences in the starting conditions lead to small differences in the later behaviour of the system is said to be linear; but a system in which small differences in starting conditions lead to big differences further down the road is said to be 'sensitive to the initial conditions', and is non-linear. The French mathematician Henri Poincaré summed up the situation as early as 1908 in his book *Science et Méthode*:

A very small cause that escapes our notice determines a considerable effect that we cannot fail to see, and then we say that the effect is due to chance. If we knew exactly the laws of nature and the situation of the universe at the initial moment, we could predict exactly the situation of that same universe at a succeeding moment. But even if it were the case that the natural laws had no longer any secret for us, we could still only know the initial situation *approximately*. If that enabled us to predict the succeeding situation with the *same approximation*, that is all we require, and we should say the phenomenon had been predicted, that it is governed by laws. But it is not always so; it may happen that small differences in the initial conditions produce very great ones in the final phenomena. A small error in the former will produce an enormous error in the latter. Prediction becomes impossible.

A very simple example makes the point. At the top of the Rocky Mountains of North America, there is a watershed, along a line called the continental divide, which marks the geographical boundary between east and west. Rain that falls on the east of the line flows away, eventually, to the Gulf of Mexico or the Atlantic Ocean; rain that falls to the west of the line flows into the Pacific. Right on the line, there must be places where a difference of less than a centimetre in the position where a raindrop falls determines its fate. Two raindrops falling from the same cloud at the same time may land less than a centimetre apart. One ends up in the Atlantic, the other thousands of miles away in the Pacific. The fate of the raindrop is sensitive to the initial conditions. But there's more. The oceans to east and west seem to attract the raindrops, and the concept of an attractor, linked with ideas about equilibrium, is the other half of the story of chaos.

An equally simple but more homely example of an attractor can be seen by rolling a marble into a round-bottomed mixing bowl. After a few ups and downs, the marble settles at the bottom of the bowl, in equilibrium in a state that corresponds to the minimum energy of that system. This state is an attractor for the system. But there may not be a unique point associated with the attractor. Take the same marble and try to balance it on the top of a pointed sombrero. It will roll off and end up somewhere in the valley made by the upturned brim, but all points around that valley correspond to the same minimum energy state (this is known in the trade as the 'Mexican Hat' potential). The whole valley is a single attractor.

In these simple examples, we are looking at systems that end up in equilibrium, with nothing changing. This is linked to the idea of entropy, which is a measure of the amount of order in a system, with increasing disorder corresponding to increasing entropy. The natural tendency of closed systems – ones which are cut off from the outside world – is for entropy to increase as things get more disordered. In the classic example, if you have a box divided into two halves, one full of gas and the other empty, then remove the partition, gas spreads out to fill the whole box evenly. There is less order, because there is no longer any difference between the two halves of the box. You may have come across the idea of entropy from this kind of example, and learned that systems are attracted to states of maximum entropy. But in the real world, there is no such thing as a completely isolated system. There is always some contact with the outside world, and this changes things significantly.

If two containers are each filled with a mixture of two gases (the classic version of the experiment uses hydrogen and hydrogen sulphide) and joined together by a narrow pipe, there will be a uniform mixture of gases in each container, at the maximum entropy for the system. But if one of the containers is kept at a slightly higher temperature than the other, the lighter molecules (in this case, hydrogen) will concentrate in the hotter container, and the heavier molecules (in this case, hydrogen sulphide) will concentrate in the cooler container. Order has been produced, and entropy has decreased. A very small deviation from equilibrium can completely change the behaviour of a system, and in general a system that is close to equilibrium but not actually in equilibrium will be attracted to a state in which the rate at which entropy is changing is a minimum. Put more simply, interesting things only happen close to equilibrium and when there is a flow of energy through a system. Exactly at equilibrium, nothing changes. Far away from equilibrium, everything changes all the time in a messy fashion – chaos. It is no coincidence that we live on a planet bathed in a flow of energy from the Sun and full of interesting things, including ourselves. Life exists on the *edge* of chaos.

The transition from an ordered system in which nothing significant happens through a state where interesting complications occur and on to a state of chaos can be seen in a real-world example. In a gently flowing river where there is just one rock sticking out above the surface, the water divides around the rock and joins up smoothly on the other side. We can monitor the flow of water by dropping little chips of wood upstream and following them down past the rock. If the flow of water gradually increases, perhaps because of heavy rain upstream, the pattern changes. At first, little whirlpools, or vortices, form just downstream from the rock. They stay in the same place, and chips of wood trapped in them go round and round repeatedly. The vortices are a kind of attractor. But as the flow of water continues to increase, the vortices get detached from the rock, and are carried downstream, holding their shape for a time before they get dissolved into the general flow of the water. New vortices form in their place and are carried off in their turn. But as the flow increases even more, the region behind the rock where vortices form and survive gets smaller and smaller. Eventually, even the water immediately behind the rock becomes choppy and moves in an irregular, chaotic way, and in a really chaotic system there are no attractors. And all this has happened because just one thing has changed – the rate at which the water is flowing. The same system behaves in very different ways if there is a change in just one thing – in terms of the mathematical modelling of a system, a change in just one number. Which brings us to that butterfly and the effect of its flapping wings.

The story goes back to 1959, when Edward Lorenz was working on computer modelling of the atmosphere as a step towards computer weather forecasting. He used the idea that a set of equations describing the state of the weather could be 'run' in a computer to predict the state days or weeks ahead, and he hoped that these equations would show that some weather patterns would be particularly stable, so they would be easy to predict. We would now call these stable states attractors. Of course, the computers he had to work with were far less powerful than those we have today, but he was not actually trying to forecast the weather in the real world, just test how the idea worked on a small scale (what scientists call a 'toy model'). The input to the model was simply a list of numbers, representing things like temperature and pressure, and the output was a corresponding list which could then be turned into a mini 'forecast'.

This 'butterfly' diagram represents the possible futures of a
system (such as the weather) balanced on the dividing line
between two 'states'. A tiny nudge will send it into one or the other
of the two whirlpools. It is 'sensitive to initial conditions'.
Science Photo Library

To his surprise, when Lorenz ran the model twice with what he thought was the same set of input data, he got two wildly different forecasts. It turned out that the difference was caused by a tiny change in the input numbers; the first time he had used six significant figures (for example, 0.506127) and the second time he had as a shortcut simply typed in numbers with three significant figures (in this case, 0.506). The typical difference in each of the numbers from one run to the next was only one quarter of one tenth of one per cent, but it changed the forecast completely. If the real atmosphere were always as sensitive as this to initial conditions, it would make computer weather forecasting a hopeless task. But it gradually became clear what is going on, and this explains why weather forecasters today are sometimes confident in their forecasts, and sometimes prefer to hedge their bets. The way to picture what is going on is in terms of a kind of imaginary landscape, which physicists call phase space.

Phase space is like a real landscape of a rolling countryside, with hills, valleys, high mountains and deep potholes. Each point on the landscape corresponds to a set of physical properties belonging to the system being investigated. So if we are representing the atmosphere, a single point on the landscape doesn't simply correspond to a single property such as temperature, but to a particular combination of temperature, pressure and other properties. A mountain peak represents an extremely unlikely state of the atmosphere, while a deep pothole represents a very attractive state. Starting a computer model running corresponds to pouring water on to the landscape at a particular point. Obeying the equations fed in to the model, the water flows downhill, and is attracted to deep pools.

These are the most likely states of the system. But the water may have a choice of paths to follow, like the raindrop falling on top of the Rockies, so where it ends up may be sensitive to the initial conditions. A slight shift in the starting point may lead to a big shift in where it ends up. The paths that can be followed in this way are called trajectories, and the points along the trajectory represent the forecast at different times in the future. A typical trajectory in phase space will settle down in one of the pools, circling round and round like the water in vortices formed behind a rock in a flowing river. But if there are two pools separated by a shallow ridge, like a sand bar, the trajectory may occasionally (and unpredictably) cross over and start circling round the other pool. It has flipped from one attractor to another. In the real world, this corresponds to a change in the atmosphere from one state to another.

What meteorologists have found in the decades since Lorenz made his discovery is that sometimes the weather is sensitive to starting conditions and sometimes it is not. These days, meteorologists don't just do one simulation starting with the numbers corresponding to today's weather in an attempt to forecast the weather for days ahead. Instead, they run the same simulation several times with slightly different starting conditions. Sometimes, these runs all give roughly the same answer. The trajectories are all heading into the same pool and circling round in it. But some days the runs give a variety of different answers, as happened to Lorenz in 1959. In that case, accurate forecasting is impossible. Not through any fault of the meteorologists or their models, but because that's the way the world is; the atmosphere at that time is in a state sensitive to initial conditions.

Which is where the butterfly comes in. Lorenz introduced it into the discussion at a meeting in Washington, DC in 1972, where he asked 'Does the Flap of a Butterfly's Wings in Brazil Set off a Tornado in Texas?' Although this is a particularly bad example, because Brazil and Texas are in different hemispheres and weather systems either side of the equator have little influence on each other, his point is that on those occasions when the weather is sensitive to initial conditions, the tiniest change can affect which way the trajectories move through phase space. A better version of the image would be if the weather systems over the tropical North Atlantic were poised on a sandbar in phase space, so that the flap of a butterfly's wings in Senegal produced an influence which tipped it towards one attractor (make a hurricane) or another (do not make a hurricane). But this was never intended as a serious example, and it certainly does not imply that small influences like flapping butterfly wings *force* big systems one way or the other. They simply provide the proverbial last straw.

There is, though, a rather more disturbing version of all this. Some climatologists suggest that among several possible stable states (attractors) for the Earth's overall climate, one corresponds to the conditions we have been used to for the past few thousand years, another corresponds to the ice age that preceded this interval, and a third corresponds to a hot state about 6°C warmer on average than today. The standard computer models of the global warming now happening as a result of the buildup of carbon dioxide in the atmosphere suggest a steady increase in temperature by a little more than 3°C as the amount of carbon dioxide increases to about twice the concentration before the industrial revolution. But Jim

Lovelock, the originator of Gaia theory, suggests that a flip into the hot state could occur before the end of the twenty-first century. The analogy would be that the two states are attractors in phase space separated by a small sand bar, and that by increasing the temperature of the globe we are forcing the trajectories to circle higher and higher up the side of one pool, until suddenly they cross over into the other one. If he is right, there is even less time to take action on global warming than most people think.

Leaving those gloomy prognostications aside, however, what can chaos theory tell us about the way the world works? What has happened to the Newtonian clockwork predictability of the Universe?

There's some good news (for us) and some bad news (for the clockwork universe idea). Popular accounts of chaos theory have led to some alarming speculations jumping off from the idea that it means everything is unstable and that because the Solar System is technically in a chaotic state, then as a result of some tiny perturbation like the passage of a comet nearby, the Earth might suddenly switch from its present orbit into another one, or plunge into the Sun. But there are degrees of chaos. That kind of chaos does apply to small objects, like asteroids, under the influence of large objects, like Jupiter and the Sun. But the Earth's orbit is only chaotic within certain limits. Modern computers are able to get round the three-body problem to a large extent using the step-by-step iteration technique, and they have been used to calculate how the orbit of the Earth is likely to change over the next few hundred million years. In the usual way, the calculation is carried out many times with slightly different starting conditions to see if this affects the outcome. The models tell us that there is only a tiny chance of any drastic change in the orbits of the eight main planets of the Solar System over billions of years – essentially, until the Sun dies. But the orbit of the Earth is sensitive to initial conditions in a limited way. In one example, changing the position of the Earth in its orbit at the start of the calculation by 5 metres does not change the final position at the end of the calculation by 5 metres. The 'error' grows until after a simulated 100 million years the model cannot say exactly where in its orbit the Earth is. All the model tells us is that the Earth is somewhere in its orbit around the Sun, which is at least reassuring for us. The whole orbit is an attractor, like the lowest energy trough of the Mexican Hat potential.

You might think, though, that all this is only a problem because we can never say *precisely* where the Earth is in its orbit, or *exactly* what the temperature, pressure and so on are at any particular point in the atmosphere. Surely if we knew these things to enough decimal places the uncertainty would disappear and everything would be predictable, just as Newton thought? And surely in some sense the Universe itself knows where everything is, so it must be deterministic? Surprisingly, the answer is a qualified 'no'.

The problem is that there are not enough decimal places. Even the Greeks knew about the problem, in a slightly different form. It has to do with the nature of numbers. There are three kinds of numbers. Integers – 1, 2, 3 and so on – are easy to understand and work with. Another family can be described in terms of the ratio of two integers, numbers like ½, ¾, and so on. These are called rational numbers (from ratio), and are also fairly easy to manipulate and work with. But the Greeks were well aware of the existence of numbers which cannot be written as ratios in this way, and are called irrational. The most important to them, and the most familiar to us, is pi (π), the ratio of a circle's circumference to its diameter. As a rough approximation, we can use a rational number like $^{22}/_7$ in our calculations, but this is *only* an approximation, as we can see if we modernise things by bringing in decimals. Using various calculation techniques, and computerised number crunching, π has been measured to millions of decimal places, and it starts off: 3.141592 65358979323846264338279.

The number $^{22}/_7$, a relatively crude approximation, starts off 3.142857, so it is already incorrect by the third decimal place. But the important point about an irrational number in decimal terms is that the pattern of numbers never repeats. The number $^1/_3$, expressed as a decimal, would be 0.333333… with the '3s' going on for ever. But as it repeats, you can specify this as a simple rule, in this case 'keep on writing 3s'. All rational numbers can be expressed by such rules, or algorithms. But to specify π, or any irrational number, precisely, there is no algorithm; you would need an infinite string of numbers, which would require a computer with infinite memory. And this is just for one number – as it happens, a vital number in calculating the orbit of the Earth around the Sun. Even worse, it turns out that most numbers are irrational. This compounds an already insurmountable problem in specifying even the position of a single point on a line precisely. Suppose that the position of that point is $^1/_\pi$ along the line between two points A and B. You can never express this exactly in mathematical terms. You can express it to as many decimal places as you like, but if the kind of chaos Lorenz discovered is at work, it may be that the next decimal digit, the one you are ignoring, alters whatever it is you are trying to calculate dramatically.

A computer with infinite memory would be required to specify the state of a *single* particle in the Universe. This means that the only system that can simulate the Universe perfectly is – the Universe. Improbably, even if everything is purely deterministic and ticking away like a cosmic clock, there is no way to predict the future precisely, the Universe itself is as ignorant about the future as we are, and for all practical purposes free will exists. Little things really do mean a lot.

All Complex Life on Earth Today is Descended From a Single Cell

There are three kinds of life on Earth, each different from the others at the fundamental level of the cell. The kind we see all around us – trees, people, mushrooms, sea snakes, you name it – is all built up from complex cells which have an inner core, the nucleus, containing the DNA which carries the instructions for life, surrounded by a bag of jelly in which interesting chemistry takes place, all held together inside a cell wall. These are called eukaryotic cells. And all of that complex life is descended from a single cell formed by the merger of two simpler cells a couple of billion years ago.

I told you this once already, in the title of this section. But it is such an improbable suggestion and it is so important that I shall spell it out a third time, for as the Bellman says in Lewis Carroll's *The Hunting of the Snark*: What I tell you three times is true. *All complex life on Earth today, including you and me and a banana, is descended from a single cell*, not in the sense that each of us is descended from a single cell formed at conception, but from one cell formed by a single act of cosmic conception some 2 billion years ago. All plants, all fungi, all animals, all algae – all descended from one single cell. As evolutionary biologists are fond of pointing out, there is no obvious difference between the cell of a mushroom and one of your cells. They both operate in the same way, translating instructions coded in DNA to make proteins and so on, even though the organisms they are part of have completely different lifestyles (unless you have some very peculiar habits).

This discovery is so astonishing that it makes related discoveries that are astonishing in their own right seem almost mundane. But they are not. The first surprise is that there are indeed two other kinds of cell, collectively called prokaryotes, both of which lack the central nucleus that is the hallmark of eukaryotes. But the distinction between these two kinds of single-celled organisms (for that is what they are) did not begin to become clear until the 1970s. Before that time, all prokaryotes were classed as bacteria, although the existence of some unusual species of bacteria was recognised. As techniques for studying the genetic material of cells developed, many of these unusual 'bacteria' were grouped together in a classification scheme and dubbed archaebacteria, because they were thought to be older than bacteria and in some sense their ancestors; but when it was realised that the origins of bacteria go back just as far, the second part of the name was dropped, and they are now known simply as archaea. Which doesn't entirely resolve the confusion, because archaea and bacteria are indeed shown by DNA and RNA analysis to be equally old. Which points to yet another surprise.

The earliest evidence for life on Earth comes from rocks 3.8 billion years old in the south-west of Greenland. These are chemical 'signatures' of life rather than fossils, but by 3.2 billion years ago, life forms in what is now Australia were leaving genuine fossil traces. In round terms, life got started on Earth about four billion years ago, only half a billion years after the planet formed. And it did so *twice*.

So much has become clear as the inner workings of archaea and bacteria have been studied in detail. At the largest scale (for a cell), archaea and bacteria have less than a third of their genes in common. At a more detailed level, the structure of their cell walls is different, and most significantly of all the evidence, the way in which they copy DNA when dividing to make new cells is different. They both use the same genetic code, but they copy it in different ways. These are very profound differences, and to imagine that both kinds of cell could have evolved from a common ancestor 'defies logic', in the words of biochemist Nick Lane. The two forms of life must have arisen separately, but out of the same kind of chemical 'soup', which would explain their similarities. The suggestion made by Lane and others is that this could have happened near hot vents in the sea floor when the Earth was young, with energy from these vents encouraging chemical processes that built up substances as complex as proteins and RNA before cells formed – a flow of energy allowing entropy to 'run backwards', as mentioned earlier. There are other suggestions – I discuss one in *Seven Pillars of Science* – and it is likely that we will never know what happened when life got started on planet Earth. We will also never know if the trick actually happened more than twice; there could have been other primordial prokaryotes that have left no descendants today. But we do know that from about 4 billion years ago right up to the present day there have been two separate forms of single-celled life on Earth. This makes both of them extremely successful survivors, but since we are all less familiar with the concept of archaea than bacteria, it is worth spelling out just how successful the former have been.

Typical bacterial cells are between about 0.2 and 2.0 micrometres in diameter, although some are long and thin, and archaea are much the same size. The first archaea to be identified live in extreme environments, such as hot springs and salt lakes, where no other organisms can survive. But they are now known to live almost everywhere, and are particularly common in the oceans, where they make up about a fifth of all microbial cells. The archaea in plankton are among the most abundant organisms on Earth. The versatility of archaea has made them vital in many roles in the environment, including carbon fixation and the nitrogen cycle. They are also part of the 'internal environment' of eukaryotic life forms – the 'microbiota' – and in people they are found in the gut, the mouth, and on our skin, all places where they are involved in many processes which help to keep our bodies ticking over. Unlike bacteria, however, there are no known archaea which cause disease. Indeed, archaea seem very good at getting along with other forms of life. Many of them are so-called mutualists, which form a mutually beneficial association with other organisms without harming each other, and others are commensals, benefiting from an association either directly or indirectly but neither helping the companion nor doing any harm. The classic example of a mutualist is an archaean called *Methanobrevibacter smithii*, which makes up about 10 per cent of all the prokaryotes in the human gut and interacts with other microbes to aid digestion. They are also found in other species, and as their name suggests, their activity produces methane. Even more significantly, though, the propensity of archaea for mutualism and commensalism suggests something else – why we are here. For more than half the time that life has existed on Earth there were no eukaryotic organisms around to contest the title of most successful form of life on Earth. Then, something happened.

You shouldn't get the idea that life was peaceful until our kind of life came along. Even at the level of single cells, life forms were competing with one another for resources, and mutations were producing new species on which natural selection could operate. Life evolved. And in the competition for resources, one way for a cell to get its metaphorical hands on new supplies was to eat another cell – or possibly for two cells to merge together, pooling their resources. It is now clear that this happened at least once about 2 billion years ago, when an archaean somehow swallowed up a particular kind of bacterium, which retained some sort of independence inside the fused cell, which became the ancestor of all eukaryotes. As with the origin of cells, such events may have occurred more than once; but the similarity of the genetic material of all complex life on Earth today shows that only one such merger produced descendants that have survived.

The story of how this happened was unravelled in reverse, starting with the study of living cells today and working out how they got to be the way they are. It began with the work of the American biologist Lynn Margulis, in the late 1960s. She was particularly interested in mitochondria, which are structures (organelles), shaped rather like grains of rice, that seem to have a semi-independent existence within eukaryotic cells, and which process the energy used by cells. In everyday terms, they take the fuel (or food) coming in to the cell and allow it to combine with oxygen (burn) to release energy (respiration). These components of cells had been discovered at the end of the nineteenth century, and it had been speculated that they were really bacteria which lived in a symbiotic relationship inside the cells. There is a similar situation in plant cells, where organelles known as chloroplasts capture the energy of sunlight and convert it into chemical energy which is used to combine water and carbon dioxide to make organic matter (photosynthesis). It became clear that chloroplasts shared many properties with cyanobacteria, small organisms that live in water, and seemed to have evolved from them. As techniques for studying DNA developed, these ideas were confirmed by the discovery that mitochondria and chloroplasts have their own DNA, distinct from that of the cells they inhabit. This gave evolutionary biologists an additional tool for investigating the relationships between species, past and present. It turned out that chloroplast DNA is indeed essentially the same as cyanobacterial DNA, while mitochondrial DNA resembles that of a group of bacteria that includes (improbable though it may seem) the one that causes typhus. But these components of cells only possess some of the DNA of their ancestors, not enough to allow them to survive an independent existence outside the environment of the cell, which is largely controlled by genetic material in the nucleus.

Margulis became the leading proponent of the idea that symbiosis was a major force in the evolution of cells, an argument summed

Lynn Margulis
Getty Images

up in her book *The Origin of Eukaryotic Cells*. Her enthusiasm led her to suggest a bacterial origin for many of the structures seen inside eukaryotic cells, and the jury is still out on some of these claims. But there is no doubt at all about the origin of chloroplasts and (most importantly for us) mitochondria. They are descended from free-living ancestors that have somehow become incorporated into other cells. The importance of Margulis' work was summed up by Richard Dawkins, who said:*

> The theory that the eukaryotic cell is a symbiotic union of primitive prokaryotic cells ... is one of the great achievements of twentieth-century evolutionary biology, and I greatly admire her for it.

When Margulis started her work, it was natural for her to assume that this involved mergers between different kinds of bacteria, because the importance of archaea had not been recognised. But it is now clear that the dominant partner in the initial merger was an archaean. Eukaryotic cells derive from a merger between two entirely separate lines of cell evolution.

The leading proponent of this development from Margulis' idea is Nick Lane, who works at University College in London. The essence of his contribution to the debate is the argument that 'the singular origin of complex life might have *depended* on the acquisition

---

* See *The Third Culture: Beyond the Scientific Revolution*, John Brockman, Simon & Schuster, New York, 1995.

of mitochondria. They somehow triggered it."* This claim is partly based on the fact that although there are a few exceptions to the rule that eukaryotic cells around today possesses mitochondria, the genetic evidence shows that every eukaryotic cell is descended from ancestors with mitochondria. To see why he makes his claim, however, we need to look at exactly what mitochondria do, and how they contribute to the workings of the cell.

Cells basically run on electricity, but whereas the particles that carry electricity around in the wires in your house are negatively charged electrons, the particles that keep the cell's machinery working are positively charged protons. They are transferred from place to place by chemical processes, and just as the flow of electricity in wires is sometimes likened to the flow of water in a pipe, so the transfer of protons from one place to another is sometimes likened to the action of a pump pushing them along. Mitochondria have a double membrane (an evolutionary development from their own original cell walls), and the inner membrane is ruffled into many folds, which means it has a large surface area packed into a small space. This surface is the key location on which chemistry involving the transfer of protons and the release of energy can take place. The chemical reactions involve an energy-carrying molecule called adenosine triphosphate (ATP), but I don't plan to go into details of the chemistry here. What matters is that one set of reactions stores energy (obtained from food) by pumping protons across the membrane in one direction, and then when the

---

* See *The Vital Question*, Profile, London, 2015. The ideas Lane describes build from work he carried out with his colleague Bill Martin.

cell needs energy it is produced by letting protons flow back the other way, like water flowing past a mill wheel and making it turn. The combination of these properties means that the mitochondria can supply a lot of energy, anywhere in the cell, whenever it is needed. And (in my opinion equally crucially) they can do so in a steady way. Prokaryotic cells control their energy supply in much the same way, chemically speaking, but all the action takes place near their cell walls, which do not have as complex a structure as mitochondrial membranes, and cannot take the energy where it is needed. In order to use the energy, prokaryotes have to have their genetic material close to the power supply, around the edges of the cell. And because genes are needed to control the workings of the machinery everywhere inside the cell, this can involve having copies of the genes in many places around the rim, which is wasteful of resources.

Cells of complex organisms use a lot of energy. The key process that keeps the cells working – keeps them alive – is the translation of instructions in the genetic code stored in DNA into proteins that do the work of the cell and provide the structure of a body. This uses three-quarters of a typical cell's energy 'budget', whether it is a prokaryote or a eukaryote. The more genes there are, the more complex an organism can be. But the number of genes is limited by the availability of energy. And remember that every time a cell divides and replicates, the entire genome has to be copied to provide a set of instructions for each daughter cell, which also requires energy. In a typical bacterium, there are about 5,000 distinct genes. But in the *smallest* eukaryote there are around 20,000 genes. An average eukaryotic cell has 200,000 more genes than a prokaryotic cell. The difference is entirely due to the availability of energy – thanks to the presence of mitochondria. On the rough and ready assumption that each gene needs the same amount of energy, a eukaryotic cell has 200,000 times more energy available than a prokaryotic cell, and it is delivered wherever, and whenever, the cell needs it.

This brings many benefits. First, it makes it possible for the cell to make extra copies of genes, something which will inevitably happen as more energy is available; we can imagine the copying mechanisms in the first eukaryotes repeatedly going about their work, like the magic brooms carrying buckets of water in the Disney movie *Fantasia*, because there is no way to tell them to stop. This copying is not always perfect, so there will be occasional copying mistakes – mutations – which provide the raw material for new versions of genes, and eventually new genes, to be produced by natural selection. Each extra gene requires more energy, but with mitochondria to hand that is not a problem. The availability of more energy speeds up evolution. Secondly, as the genome no longer has to go to the source of energy, the genes can be packed away at the heart of the cell, in the nucleus, keeping them out of harm's way and leaving space for the cell machinery to do its work. This development must itself have been a result of the evolution encouraged by the availability of energy, but it is unlikely that we will ever know exactly how it happened. Meanwhile, the ancestors of the mitochondria lost the genes that enabled them to survive outside the environment of their host cell, but kept the ones involved in processing energy. They too evolved, to be more efficient providers of that energy. The end product is the eukaryotic cell we know today. But what was the initial 'product'? And how can we be sure that there was indeed only one cell that became the ancestor of all eukaryotic organisms on Earth today?

Lane's argument (which is not accepted by everyone, but looks good to me) is that about 2 billion years ago (long before life moved onto the land) there was a population of archaea living in the ocean and a population of bacteria living alongside them. The two stayed in close proximity because there were mutual benefits, or at least benefits to one side of the relationship. He suggests that one of them may have been feeding off the waste products of the other, but offers this only as a guess. If something like this were going on, the closer the beneficiaries – let's say, bacteria – got to the source of their food – the archaea – the better they would thrive. At some point, at least once, a bacterium snuggled up so close to an archaean that it got inside, and (improbably) wasn't eaten. The invader was tolerated because it did no harm, and it thrived because whatever waste products it needed were available all around it. As it became more dependent on that source, it lost the genes it no longer needed, and became simply a powerhouse.

This is what is sometimes called a 'just so' story, after the book by Rudyard Kipling. It might have happened like that, but the story is really a parable to set us thinking about the possibilities. What matters is that an archaean and a bacterium did get together, and however many times this happened, only one of those unions has left descendants alive today. All distinctly eukaryotic features evolved after this union took place. These features are common to all eukaryotes – even what Lane refers to as 'picoeukaryotes', which are 'tiny but perfectly formed cells … as small as bacteria yet still featuring a scaled-down nucleus and midget mitochondria'. In all eukaryotic cells, the nucleus is surrounded by a particular kind of double membrane, they all have straight (well, linear) chromosomes (prokaryotes have their genes strung round loops of DNA), they use the same chemical processes to operate the machinery of the cell, and they all reproduce sexually, which plays a major role in evolution.* There is no need to labour the point, since with modern sequencing techniques it is possible to analyse their DNA and see directly how closely related they are. The evidence is compelling. But before I move on, I want to point out an often overlooked feature of the role of mitochondria.

Mitochondria don't just supply energy to the cell. They do so in a controlled way. Which brings me back to the discussion of chaos and complexity. Life exists close to, but not at, equilibrium, feeding off a flow of energy. In the eukaryotic cell, that flow is controlled by mitochondria. If the flow is too slow (like the water flowing smoothly past a rock in a river), we get closer to equilibrium, and nothing interesting happens – biochemistry stops and the cell dies. If the flow is too fast (like a torrent smashing against the rock), we have chaos and the smooth biochemical running of the cell is disrupted. It dies. We, along with all eukaryotic organisms, are utterly dependent on mitochondria maintaining the knife-edge balance between two forms of death.

What does all this tell us about the prospects of life existing elsewhere in the Universe? The good news is that if two different forms of cellular life got going on Earth almost as soon as the planet had cooled, the chances of life existing on other planets must be

---

* See John Gribbin and Jeremy Cherfas, *The Mating Game*, Penguin, London, 2001.

high. The bad news is that if it took 2 billion years before a chance encounter between two prokaryotic cells led to the beginning of eukaryotic life, the chances of such complex life existing on other planets must be very small. And it may be even smaller than it looks at first sight. That chance encounter took place between two different forms of prokaryotic life, where each brought their own distinctive packages of genes to the merger. This immediately gave the first eukaryotes a more complex genome and the raw material for evolutionary processes to operate on. If you need two different kinds of simple cell to merge in order to produce the kind of complex life that eventually led to us, the chances of there being life forms as complex as us on other planets becomes vanishingly small. Even on a cosmic scale, our existence is highly improbable. And even complexity doesn't necessarily imply the evolution of our kind of intelligence. After 2 billion years of eukaryotic evolution, it took an improbable set of circumstances to turn an African tree-ape into *Homo sapiens*.

Ice Age Rhythms and Human Evolution: People of the Ice

The evolution of life on Earth has always been influenced by environmental and climatic changes. One of the most dramatic of these was the event – or series of events – that happened about 65 million years ago and brought an end to the reign of the dinosaurs. This almost certainly involved the impact of a largish meteorite with our planet, although other factors may also have played a part. Since the death of the dinosaurs led to the rise of the mammals, this is a good place to pick up the story of our own origins. But to put this in perspective, 65 million is just over 3 per cent of 2 billion. Eukaryotic life had been evolving for 97 per cent of the time from its origin to ourselves before that meteorite struck.

A variety of geological evidence shows that during the 60 million years or so that followed the death of the dinosaurs, while the mammals were diversifying and filling many ecological niches left vacant by their predecessors, the temperature of the Earth slowly and unevenly declined, as a result of the way the continents were moving around on the surface of the globe, changing the way sunlight was absorbed and reflected, and altering the flow of ocean currents. But by about 4 million years ago a tipping point had been reached.

The evidence suggests that 65 million years ago there were no large ice sheets on Earth, although there may have been seasonal snow on mountain tops. This situation began to change about 13 million years ago, as Antarctica drifted slowly across the South Pole, and ice sheets began to form in what is now East Antarctica. By 10 million years ago, there were small glaciers on the mountains of Alaska. Around 6 million years ago, Australia and South America moved away from Antarctica, leaving a clear passage for a strong ocean current, the circumpolar flow, to surround Antarctica, keeping warmer water at bay and locking the continent into a full ice age. Things were different in the northern hemisphere, where at first warm currents flowed right up to the pole, keeping the Arctic Ocean ice-free. But there, the drifting continents were slowly sliding into the positions we know today, gradually surrounding the polar ocean and drastically reducing the flow of warm currents to the Arctic. In the south, there was a continent permanently covered by ice; but in the north, an ice-covered ocean developed, and there was large-scale glaciation over the land surrounding that ocean by about 3.6 million years ago. The world was plunged into an Ice Epoch, during which ice sheets grew and shrank, but never completely disappeared. The situation that we think of as normal, with ice over both polar regions, is extremely rare, and possibly unique in the long history of the Earth. The fact that we have different kinds of glaciation in each hemisphere is also highly improbable.

And the unusual nature of the northern ice cap makes the whole world particularly sensitive to climatic fluctuations that are key as far as human origins are concerned. It is no coincidence that our line developed during the Ice Epoch; but it wasn't cold that was the driving force, it was drought.

An ice age is also a dry age. Water that is locked up in ice sheets would otherwise be in the sea, so when there is more ice on land, the sea level is lower.* Just under 6 million years ago, the ice sheets over Antarctica were several hundred metres higher than they are today, and so much water was locked up in them that sea level fell by about 50 metres (compared with today). This was too low to allow water to flow through the shallows of the Gibraltar Strait, and the Mediterranean dried up; it actually dried up and refilled repeatedly, as the ice sheets fluctuated in size. There was also desert in modern-day Austria. This desertification was linked to the cooling of the globe, because when the world is cooler there is less evaporation of moisture from the oceans, so there is less rainfall. With lower sea levels, the boundary between land and sea was further away from the interiors of continents, so what rain-bearing systems there were had a good chance of dropping their load before they even got to places like Austria. More significantly for the story of human origins, the droughts associated with ice ages also affected the forests of eastern Africa. The temperature there didn't change much as the ice sheets to the north ebbed and flowed; but the rainfall did. I shall go into why the ice sheets ebbed and flowed shortly,

---

* Floating ice has no effect on sea level because it occupies the space of the water it displaces.

but whatever the cause, what matters is that for the past few million years east Africa has been subject to a roughly rhythmic pattern of more and less rainfall – times of feast and times of famine.

I have learned to be cautious about going into too much detail about the specifics of the evolutionary line that leads to ourselves, because new evidence is still being uncovered, and the experts sometimes revise the details of the picture. But the overall picture, based on a combination of fossil evidence and DNA sequencing, does not change. I shall focus on how things evolved from the time when our ancestral line split from the ancestral lines of our nearest relations, the African apes called the gorilla and the chimpanzee.* These are all members of a group, in which we are included, classed as hominids; the term hominoid covers a larger variety of apes including our more distant cousins. In very round numbers, the split that leads to us happened between about 3.5 million and 4 million years ago, with some evidence that the line leading to gorillas split off first, and then the split between ourselves and the chimps occurred. In geological terms, this is intriguingly close to the time when Antarctica drifted over the South Pole and the climate of eastern Africa began to change. Combining evidence from a variety of sources, it is clear that over the next few million years a proto-ape species living in the forests of eastern Africa gave rise to three closely related but distinct ape lines, just at the time the climate was changing significantly. The most plausible speculation is that

* On any reasonable classification system, we would also be regarded as African apes, but as it was people who made the classification we have been put in a category of our own.

the evolutionary changes were a response to the environmental changes.

It isn't difficult to see how this could have happened. When the forests get dry, they shrink. This reduces the availability of resources and increases the competition between individuals. It's worth spelling out just what competition means, in evolutionary terms. Individual members of a species are not in competition with other species, but with each other. When lions hunt deer, the lions are competing with each other to catch prey, and the deer are competing with each other to run away. The resulting arms race leads to lions with better hunting skills, and deer with better running skills, as bad hunters starve and slow runners get eaten. In the shrinking forests, individual apes that were better at climbing, say, got more fruit, survived, and produced more offspring than their rivals. But on the edge of the forest, another option was open to the less successful climbers. They got pushed out and had to cope as best they could on the savanna, where those that were better able to cope with the new lifestyle – for example, better at walking upright – did best and left most descendants. This could explain the changes that split us from the other ape lines and made us human.

Exactly which hominid was the direct ancestor of a particular later hominid is not always clear, but the first one to be given the genus name *Homo*, *Homo habilis*, was around in east Africa by about 2.5 million years ago. *Homo habilis* was an ape that walked upright, stood about 1.2 metres tall, and had a slender build but a relatively large head with a brain capacity of 675 cubic centimetres, about half that of our modern species, *Homo sapiens*. By 1.5 million years ago, *Homo erectus* was on the scene – 1.6 metres tall, with a brain size of

925 cubic centimetres. This was the species that spread our ancestral line out of Africa and into Asia. It wasn't until 500,000 years ago that *erectus* had evolved into *Homo sapiens*, the modern human form that eventually spread to every continent on Earth.

But there was more to the environmental changes that accompanied, and probably caused, this evolution than a simple slide down into cooler and drier conditions. Curiously, the geological record shows that over the past few million years the Ice Epoch has been broken up into a repeating pattern in which the ice advances and there is a full ice age for about 100,000 years, then there is a slight warming and the ice retreats into what is called an interglacial state for about 10,000 years. All of human civilisation has developed during the most recent interglacial, but we are still in an Ice Epoch. In eastern Africa, this means that for an interval of 100,000 years or so the forests dry out and times are hard. In the heart of the forest, the successful tree-dwellers are largely unaffected and their lifestyle continues unchanged. But out on the edges of the forest there is a strong evolutionary pressure as many individuals die. The few survivors are increasingly well adapted to the conditions, but may be so reduced in numbers that they are on the edge of being wiped out. Then, there are 10,000 years or so of plenty, and the survivors go forth and multiply. Each turn of the environmental screw ratchets evolution up by another notch.

What survival characteristics will be ratcheted up in this way, in the borderland between forest and savanna? In two words, adaptability and intelligence. And of the two, adaptability is arguably more important. Some animals run faster than us, some swim better, some have more efficient claws and teeth for killing and eating meat, and some have digestive systems better suited than ours to digesting plants. But we do a little bit of everything fairly well – exactly the survival traits needed when resources are scarce and there is fierce competition for them. Intelligence, especially the ability to work out in advance where the next meal is coming from, is the icing on the cake. None of this would have been needed* if there had been no Ice Epoch and there had been plenty of lush forest full of resources. Conversely, if there had been no let-up in the drought, the population of apes driven to the fringes of the forest might have been wiped out before these traits could evolve. It is the peculiar rhythm of the ice ages that has made us human.

This is not just speculation, because we have hard evidence of this pattern of climatic change. The broad picture is revealed by a variety of geological records, spanning millions of years, but the clinching evidence of what has been going on comes from the detailed record of the past million years or so. The details are provided by isotopes of elements such as carbon and oxygen, trapped in bubbles of air in Antarctic ice, or in the form of carbonates in the shells of long-dead creatures in the mud of the sea floor. Cores drilled from the ice or mud contain samples laid down year by year, so going deeper down the core is like looking back in time, and different layers can be dated by a variety of techniques which I have no room to describe here.[†] The isotopes tell a tale because the proportions of them in the air, and therefore in the bubbles or the shells, depends on temperature. For example, oxygen-18 is heavier than

---

* By which I mean, there would have been no selection pressure for these traits to evolve.
[†] See John Imbrie and Katherine Imbrie, *Ice Ages*, Harvard University Press, 1986.

oxygen-16, so water (H_2O) that contains O-18 is harder to evaporate from the sea; and the balance between carbon isotope ratios in the carbonates of deep-sea sediments also tells researchers what the temperature was when those sediments were being formed. Similar effects are used to reveal temperatures of the past from ice cores. The results show a complicated pattern of changes over the past couple of million years, but this can be unravelled (another job for power spectrum analysis) to reveal that it is dominated by a mixture of three repeating cycles and some minor components. It is these cycles that cause the pattern of ice ages and interglacials. But the discovery of these rhythms, in the mid-1970s, did not come as a surprise, because this pattern of ice ages had been predicted before there was any geological evidence for it, and before anybody knew anything about the details of human origins.

The prediction developed from work by a Scottish scientist, James Croll, in the nineteenth century; but it was worked out in painstaking detail by the Serbian Milutin Milankovitch, carrying out enormously lengthy calculations literally using pen and paper, mostly while he was a prisoner of war in Hungary during the First World War. The result is sometimes referred to as the astronomical theory of ice ages, but more informally as the Milankovitch Model. It depends entirely on the unusual geography of the globe today, with an ice-covered Arctic Ocean almost completely surrounded by land. Because of this configuration, every winter snow falls on land at high latitudes. At present, during an interglacial, every summer the snow melts. But what would happen if it didn't melt? Snow is white, and highly reflective. By reflecting away the Sun's heat, it would cool the globe, and this happens no matter how thin

Milutin Milankovitch
Science Photo Library

the snow cover is. In the following winter, which starts off cooler than the one before, more snow falls, both on top of the snow from previous years and further to the south. In a short span of time, geologically speaking, you have an ice sheet which grows upward and outward. There is a positive feedback which will maintain the ice age until something significant changes. None of this can happen over the sea, where the snows of winter melt on contact with water that is warmer than freezing point. The relevant question is not why we have ice ages. Given the present geography of the globe, the natural state of the northern hemisphere is to be in an ice age – Antarctica is permanently ice-covered anyway. The question which needs to be addressed is why we ever have interglacials. Which is where the calculations of Milankovitch and his successors come in.

What matters is not how cold the winters are, but how warm the summers are. An ice age only ends when there is a run of warm summers which melt the edges of the ice back, revealing dark earth which absorbs more solar heat and speeds the melting in another feedback process. The surprise, to non-astronomers, is that the balance of the seasons does change in this way, and it does so because of changes in the orbit of the Earth as it moves around the Sun, and the way it wobbles on its axis as it orbits. This is what Milankovitch spent years calculating by hand, decades before the advent of electric computers.

It will come as no surprise to you to learn that there are three main components to these changes. The longest cycle concerns the orbit itself, which because of the gravitational influences of other objects in the Solar System changes from being slightly more elliptical to being more nearly circular and back again roughly every 100,000 years. At present, the orbit is very nearly circular (the eccentricity is close to zero), but a few score thousand years ago it was relatively elongated, with an eccentricity of about 6 per cent. Another effect, called the precession of the equinoxes, results from the way the Earth wobbles like a spinning top. An imaginary line joining the North Pole to the South Pole is not perpendicular to a line joining the centre of the Earth to the centre of the Sun, but is tilted at about 23.4 degrees out of the vertical. It is this tilt, as I mentioned earlier, that gives us the cycle of the seasons; during the part of the Earth's orbit where the North Pole is leaning towards the Sun, it is summer in the north. Six months later, it is winter in the north. And always, of course, it is the opposite in the south. Over a single orbit, the North Pole always 'points' to the same part of the sky (to the same place in the background of stars); but over a cycle roughly 20,000 years long it traces out a slow circle on the sky.

But this isn't all it does. On a longer timescale, roughly 41,000 years, the tilt itself changes, nodding up and down over a range from 24.4 degrees (most tilted) to 21.8 degrees (most upright). The present tilt is roughly halfway between these extremes, and has been decreasing for the past 10,000 years. This means that for the past 10,000 years the contrast between the seasons has been getting less. It is not a coincidence that the most recent ice age ended, and the present interglacial began, when the tilt was more extreme and there was a bigger contrast between the seasons. Although the total heat received from the Sun over a whole year is always the same, what matters is how hot the northern hemisphere summers are, regardless of how cold the winters are.

Crunching all the numbers in modern computers, and including some minor effects, there is a very close match between the calculations of the amount of heat received by the northern hemisphere in summer and the pattern of ice ages and interglacials revealed by the ice cores and deep-sea cores. The astronomical theory of ice ages is correct.

But this is not quite the end of the story. What is it that controls the tilt and wobble of the Earth? The Moon. Without the stabilising influence of the Moon, as I mentioned earlier, the Earth's tilt could vary by as much as 85 degrees. In such a situation, the extreme fluctuations in climate would make it impossible for life forms like us to evolve. It is thanks to the Moon that we have the rhythms of the Milankovitch Model which have made human beings out of forest apes. Which is a suitably improbable note on which to leave you.

FURTHER READING

Easy stuff

Marcia Bartusiak, *Einstein's Unfinished Symphony*, Yale University Press, 2017

John Gribbin, *Deep Simplicity*, Penguin, London, 2005

John Imbrie and Katherine Imbrie, *Ice Ages*, Harvard University Press, 1986

Lawrence Krauss, *A Universe From Nothing*, Free Press, New York, 2012

James Lovelock, *The Revenge of Gaia*, Allen Lane, London, 2006

Not so easy stuff

Nick Lane, *The Vital Question*, Profile, London, 2015

Richard Westfall, *Never at Rest*, Cambridge University Press, 1983

Hard stuff

Charles Misner, Kip Thorne and John Wheeler, *Gravitation*, Princeton University Press, 2017

Thanu Padmanabhan, *After the First Three Minutes*, Cambridge University Press, 1998

Fictional stuff

John Gribbin and Marcus Chown, *Double Planet*, Gollancz, London, 1988

Rudyard Kipling, *Just So Stories*, Wordsworth Children's Classics, London, 1993

ALSO AVAILABLE BY JOHN GRIBBIN

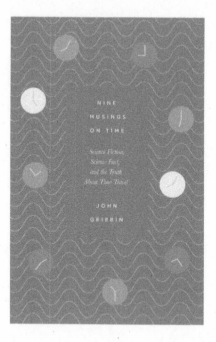

NINE MUSINGS ON TIME

Science Fiction, Science Fact, and the Truth About Time Travel

Time travel is a familiar theme of science fiction, but is it really possible? Surprisingly, time travel is not forbidden by the laws of physics – and John Gribbin argues that if it is not impossible then it must be possible.

Gribbin brilliantly illustrates the possibilities of time travel by comparing familiar themes from science fiction with their real-world scientific counterparts, including Einstein's theories of relativity, black holes, quantum physics, and the multiverse, illuminated by examples from the fictional tales of Robert Heinlein, Isaac Asimov, Gregory Benford, Carl Sagan and others.

The result is an entertaining guide to some deep mysteries of the Universe which may leave you wondering whether time actually passes at all, and if it does, whether we are moving forwards or backwards. A must-read for science fiction fans and anyone intrigued by deep science.

ISBN 978-178578-917-5

£10.99